SpringerBriefs in Materials

The SpringerBriefs Series in Materials presents highly relevant, concise mono-graphs on a wide range of topics covering fundamental advances and new applications in the field. Areas of interest include topical information on innovative, structural and functional materials and composites as well as fundamental principles, physical properties, materials theory and design. SpringerBriefs present succinct summaries of cutting-edge research and practical applications across a wide spectrum of fields. Featuring compact volumes of 50 to 125 pages, the series covers a range of content from professional to academic. Typical topics might include

- A timely report of state-of-the art analytical techniques
- A bridge between new research results, as published in journal articles, and a contextual literature review
- A snapshot of a hot or emerging topic
- An in-depth case study or clinical example
- A presentation of core concepts that students must understand in order to make independent contributions

Briefs are characterized by fast, global electronic dissemination, standard publishing contracts, standardized manuscript preparation and formatting guideli-nes, and expedited production schedules.

More information about this series at http://www.springer.com/series/10111

Ajay K. Singh

Microbially Induced Corrosion and its Mitigation

 Springer

Ajay K. Singh
Indian Institute of Technology Roorkee
Roorkee, India

ISSN 2192-1091   ISSN 2192-1105   (electronic)
SpringerBriefs in Materials
ISBN 978-981-15-8017-8   ISBN 978-981-15-8019-2   (eBook)
https://doi.org/10.1007/978-981-15-8019-2

This Springer imprint is published by the registered company Springer Nature Singapore Pte Ltd.
The registered company address is: 152 Beach Road, #21-01/04 Gateway East, Singapore 189721, Singapore

Dedication to—

My Late father, who within his limited means saw that all his kids acquire the best possible education, my late mother for preparing tasty meals I still miss.

My wife, RAMA, whom I love soooo much and who has happily ignored my tantrums throughout.

My daughters SHRUTI and SMRITI and my grandkids KUHU and SHAARV who are so adorable.

Late Prof. (Dr.) Yogeshwar Singh—My brother-in-law for letting me understand the hard realities of life and taking them in good stride.

Preface

How and Why the Book Came into Being

During the course of last 25 years or so, the author has come across several books on microbial induced corrosion. These books have covered a lot related to details about bacteria and its microbiological, DNA, etc., aspects, but not so much on the electrochemistry, and engineering aspect. In the literature too, one finds that not much work has been done which leads to enrich knowledge on how electrochemical aspects utilizing metabolic reactions and resultant bioproducts help in understanding the mechanism of MIC in case of different bacteria. Further, most MIC-related work has been done either by microbiologists or at the best a team of microbiologist with supporting role of a corrosion engineer. At the same time, one is encountering increasing number of cases of MIC and related failures of industrial and structural components where MIC is prevalent in an environment having a consortium of bacteria. This obviously is responsible for ever-increasing cost associated with overcoming ill aspects of machinery failure. The need, therefore, is of producing the microbiological corrosion engineers who have background of both the areas and therefore can work more effectively in overcoming industrial problems due to MIC. For this purpose, it is necessary to write a book which emphasizes equally on microbiology, electrochemistry and engineering aspects of corrosion in the presence of either a single bacterium or a consortium of bacteria.

Brief Description of the Book

The text matter in this book is covered in 6 chapters. Chapter 1 deals with introduction of corrosion as an electrochemical phenomenon, different types of corrosion and measurement and testing. Chapter 2 discusses theories and various phenomena related to MIC considering cases of some of the common bacteria. Chapter 3 dwells on the importance of economy aspects of corrosion/MIC and how its mitigation affects in redeeming a significant proportion of this extra cost. It also gives a brief idea about industries affected by MIC. Chapter 4 discusses experimental details about microbiology and corrosion aspects. Afterward, it dwells with the strategies to be adopted for controlling MIC. Chapter 5 deals with some industrial cases of MIC. And Chap. 6 covers techniques/approaches being utilized presently and likely to be utilized in the near future to mitigate MIC.

Purpose of the Book

The purpose of the book is to introduce its readers a highly interdisciplinary field of microbial corrosion engineering of great importance for the economy of a country. This field, namely 'microbial induced corrosion,' utilizes the knowledge of natural science, e.g., life science, physics, chemistry, and materials science, and engineering, e.g., biotechnology, chemical, metallurgy, and mechanical engineering.

Audience This book is meant for undergraduate/postgraduate and research students having interest in pursuing academic/research work in the area of microbial engineering, deterioration of materials due to interaction with microorganisms, microbiological induced corrosion inhibition, and similar such area bearing implication on enhancing the working life of industrial plants, manufacturing units, structures, vehicles, etc., situated in an environment swarmed by microorganisms.

Prerequisite The audience is expected to have familiarity with electrochemistry and construction materials for which background in graduate-level physics, chemistry, materials science and properties of engineering materials, namely chemical resistance, structural, mechanical, thermal, physical, design and fabrication and cost analysis is required.

Roorkee, India   Ajay K. Singh

Acknowledgements

Late Prof. N. J. Rao, Chemical Engineering at IIT Roorkee, Ex-Director of Central Pulp and Paper Research Institute, Saharanpur, and Ex-Vice–Chancellor, Jaypee University of Engineering and Technology, Guna—gem of a person, for encouraging and helping me at every step through my academic professional career.

Late Prof. Marcel Pourbaix and Ir. Antoine Pourbaix of CEBELCOR, Brusseles, Belgium, for guiding me, during my stay there in the 1990s, in mastering the potential—pH diagram, commonly known as Pourbaix diagram, a master piece technique in the field of electrochemical corrosion in understanding its mechanism under different metal–environment systems.

Professor Töre Ericsson and Prof. Lennart Häggström of Institute of Physics and Astronomy, Uppsala University, and Jan Gullman of Swedish Corrosion Institute, Stockholm, Sweden, with whom I initiated work on atmospheric corrosion from 1982 during my yearlong stay there.

Professor V. Pruthi, Department of Biotechnology, IIT Roorkee, India, in collaborating with me on microbial induced corrosion work. Professor S. P. Singh, Chemical Engineer, and Prof. Satish Kumar, Chemistry at IIT Roorkee, who were a great help, whenever needed, related to chemical engineering and chemistry aspects of corrosion.

Dr. Suman lata and Ms. Reena Sachan, my two Ph.D. students in the field of 'microbial induced corrosion,' long discussions with whom helped me a lot in understanding intricacies of microbiology which initially was totally an unknown subject to me.

Contents

Chapter 1
Introduction to Corrosion

Abstract Since this book is intended to initiate students into an interdisciplinary field of microbial induced corrosion, a glimpse of the importance of this course is provided to begin with. Further, since the reader may not be having appropriate background of corrosion, the first chapter of the book starts with the aim to let him/her become aware of fundamentals of corrosion. Accordingly, the chapter starts with basic definition of corrosion, various important chemical/electrochemical reactions, thermodynamics, and kinetic aspects related to corrosion. Then, it goes on to discuss different types of corrosion that occur under various circumstances. Finally, the chapter gives an idea about the basic aspects related to measurement and testing for quantification of extent of corrosion experienced by materials in laboratory as well as in-plant conditions. These measurements help in deciding about the suitability of material of construction for a given chemical environment, in laboratory as well as in real-life conditions.

Keywords EMF series · Galvanic series · Electrochemical theory · Electrochemical tests · Immersion tests · In-plant tests

This book covers various aspects related to corrosion, its fundamentals in terms of thermodynamics, electrochemistry, metallurgy, chemical engineering, etc., its effect on the machinery of various industries and thereby affecting their economics of operation, etc. We shall learn, while undergoing through the book, that corrosion is an adverse effect on materials due to the presence of chemicals in their vicinity at given temperature, pressure, and mechanical stressing conditions. However, in many instances, the presence of these conditions may be benign but due to the presence of microorganisms, the electrochemistry of the material–environment system may change so as to initiate and trigger an unacceptable level of attack on the machinery materials resulting into shortening of their useful life and in extreme cases resulting in explosion/accidents culminating in loss of invaluable human life. These are the examples of 'microbial induced corrosion' nicknamed as 'MIC' or 'biocorrosion.' Although cases of MIC were identified long back but due to their scant observance, lack of experimental techniques of their detection, understanding of fundamentals of corrosion as a phenomenon, and last but not the least non-availability of scientists/engineers/technologists having basic understanding of physical and chemical

© The Author(s), under exclusive license to Springer Nature Singapore Pte Ltd. 2020 1
A. K. Singh, *Microbially Induced Corrosion and its Mitigation*,
SpringerBriefs in Materials, https://doi.org/10.1007/978-981-15-8019-2_1

Table 1.1 Corrosion cost industry-wise

S. No.	Industry	Cost (billion $)
1	Gas and liquid transmission pipelines	7
2	Waterways and ports	0.3
3	Gas distribution	5
4	Drinking water and sewer systems	36
5	Ships	2.7
6	Oil and gas exploration and production	1.4
7	Petroleum refining	3.7
8	Chemical, petrochemical, and pharmaceutical product	1.7
9	Pulp and paper	6.0

science and engineering together with life science, MIC could not find much attention. These reasons can be assigned to lack of literature in the area of 'microbial induced corrosion' discussing an in-depth analysis of the phenomenon. However, in recent years, due to many of the industrial cases reporting the occurrence of MIC and researchers from interdisciplinary areas involving material/chemical/life scientists, engineers and technologists undertaking the research related to MIC one observes increased understanding of MIC and availability of large data on MIC. For beginners in this field, there is still lack of authentic textbooks in this area. The present book, therefore, is an effort in the direction of demystifying the phenomenon of 'microbial induced corrosion' which ultimately help in mitigating the adverse effects of MIC and thereby improve the economy of industries affected by MIC.

As indicated above, corrosion results in premature loss of machinery and so additional efforts are to be made related to maintenance to avoid premature failure, use of costly material of construction, e.g., stainless steel, titanium, plastics/rubbers, etc., maintenance of inventory to avoid production layoff due to unpredicted failure of machinery, loss of chemicals, human life in case of catastrophic failures, etc. This all involves money, and a part of it can be avoided if one takes suitable preventive measures. Table 1.1 gives an idea of the loss incurred due to corrosion by some industries [1]. This distribution may not be same everywhere but has to be country specific.

According to a 2002 study of National Association of Corrosion Engineers (NACE) [2], the global annual cost of corrosion is USD 276 billion, which is more than 3% of US GDP. Though the cost due to MIC has not been calculated independently according to various researchers of this field, the MIC is estimated to contribute to approximately 20–50% of this total [3]. This cost involves expenditure incurred on materials, overdesign, equipment, and services involved with repair, maintenance, and replacement only without accounting for trillions of $ gone down the drain due to environmental damage, waste of resources, loss of production and personal injury/loss of human life resulting from corrosion, loss of chemicals due to leakages as a result of corrosion damage, loss of appearance-automobiles, buildings, contamination of food and degradation of product etc. It is estimated that roughly

1/5th of this cost or ~55 billion USD may be saved by using appropriate corrosion control technology. In India, annual loss due to corrosion has been estimated at Rs. 7000 billion (~ 3–5% of GDP) equivalent to USD 113 billion [1]. Consequently, fighting corrosion is a real and very large market that has been a focus of manufacturers since steel galvanizing was popularized almost 200 years ago. After discussing, the importance of corrosion in nearly every walk of life, we now move forward to discuss few fundamentals in order to understand the basics of corrosion, in the remaining part of this chapter.

1.1 Corrosion Theory

Corrosion is defined as a phenomenon, which occurs naturally, and weakens the material or affects the appearance of a metallic surface or makes a smooth surface rough by way of several electrochemical reactions. Some of these reactions are oxidizing, and some are reducing type. Examples of some such reactions are given below:

Oxidizing Reactions:

$$Fe \rightarrow Fe^{2+} + 2e^- \tag{1.1}$$

$$Al \rightarrow Al^{3+} + 3e^- \tag{1.2}$$

$$Mg \rightarrow Mg^{2+} + 2e^- \tag{1.3}$$

$$Fe^{++} + 2H_2S \rightarrow FeS_2 + 4H^+ + 2e^- \tag{1.4}$$

For metal corrosion, first three reactions are important because they convert metal into metal ion which in turn form some compound. These compounds form in solution (Fig. 1.1) as the metal ions move into the solution while electrons remain on metal (metal being good conductor). Metal ions in solution may combine with anions to form compounds and remain stuck on metal surface. These compounds (e.g., Fe_2O_3, $FeSO_4$, etc.) are weaker and brittle than their metal counterpart, so on continuing oxidation, more and more of metal will transform into metal compounds and therefore stress on material becomes higher than its tensile strength which results in its failure. The material of construction is marked being weaker, with the replacement of metal-by-metal compound. In other words, its load-bearing capacity deteriorates due to corrosion (Fig. 1.1).

Formation of these compounds also affects the appearance as they are colored in many cases and makes the surface rough due to their random distribution on the

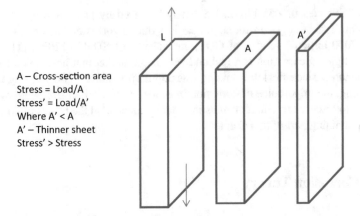

A – Cross-section area
Stress = Load/A
Stress' = Load/A'
Where A' < A
A' – Thinner sheet
Stress' > Stress

Fig. 1.1 Thinning of metal sheet leading to loss of its load-bearing capacity

surface. It is also observed that change in Gibbs free energy on oxidation of metal is negative (Table 1.2); hence, these oxidation reactions are spontaneous and so is corrosion unless some special measures are taken.

Some of the reduction reactions are:

$$O_2 + H_2O + 4e^- \rightarrow 4OH^- \tag{1.5a}$$

$$O_2 + 4H^+ + 4e^- \rightarrow 2H_2O \tag{1.5b}$$

$$2H^+ + 2e^- \rightarrow H_2 \tag{1.6}$$

$$Fe^{3+} + e^- \rightarrow Fe^{2+} \tag{1.7}$$

These reactions are due to constituents of the medium to which a metal may be exposed. Thus, O_2 remains in any aqueous solution as dissolved oxygen, H^+ ion

Table 1.2 Gibbs free energy

Metal oxidation	$\delta G°$ at 298 °K (kJ/mol)
$Al \rightarrow Al^{3+}$	−1582.0
$Cr \rightarrow Cr^{3+}$	−1045.0
$Fe \rightarrow Fe^{3+}$	−741.0
$Mg \rightarrow Mg^{2+}$	−569.6
$Zn \rightarrow Zn^{2+}$	−318.2
$Ag \rightarrow Ag^+$	−13.0
$Au \rightarrow Au^{3+}$	+163.0

concentration defines the pH of solution, etc. Further, these reactions consume electronic charge. Since no energy is being supplied, from outside, for both oxidation and reduction reactions to occur, the system should remain electrically neutral overall. In other words, number of electrons produced per unit time by oxidation reaction should be same as number of electrons utilized in reduction reactions. This is possible when **overall rate of oxidation reaction is same as overall rate of reduction reaction**. This gives us an important principle of corrosion which helps in understanding as to whether a particular change in process chemistry of the media will result in decrease or increase of corrosion.

Thus, one finds that boiler corrosion is minimized by deaerating the water meant for drawing heat.

Without deaeration, the reduction reactions will be reactions 1.5a and 1.6. While in case of deaerated water, reduction reaction will only be 1.6, there will be no significant amount of dissolved oxygen. Thus, reduction reaction rate in deaerated water will be lesser hence oxidation rate of iron (a major constituent of boiler steel). This in turn will reduce corrosion of boiler tubes. Similarly, one can understand that if a solution is made more acidic, it becomes more corrosive because of increase in reduction rate due to H_2 evolution reaction, which in turn enhances oxidation rate of metal.

We see here that both oxidation and reduction reactions involve the presence of electronic charge; hence if an oxidation reaction is occurring on a metal, that metal will become negatively charged surrounded by cations such that rate of oxidation (Fe $\rightarrow Fe^{2+} + 2e^-$) is same as rate of reduction reaction ($Fe^{2+} + 2e^- \rightarrow$ Fe). Under this situation, there is an equilibrium concentration of metal ions in the vicinity of metal. This metal then acts as electrode because of it possessing an electrical potential. This potential is known as 'electrode' potential and is also called as 'redox' (from reduction–oxidation) potential. One can see from any textbook on electrochemistry that the redox potential is related to the change in Gibbs free energy as a result of an oxidation–reduction reaction by

$$\Delta G = -nFE \tag{1.8}$$

where 'n' is oxidation state of metal ion, F is Faraday, and E is redox potential. Obviously, this potential of electrode depends upon the concentration of ions in its vicinity represented by Nernst equation

$$E = E^0 + (2.303RT/nF) \log_{10}(a_{ox}/a_{red}) \tag{1.9}$$

(The convention, in general, is to consider the potential as reduction potential.) a_{ox} is activity of oxidized species, while a_{red} is activity of reduced species. For oxidation reaction of iron (Eq. 1.1), a_{ox} is a_{Fe2+} whereas a_{red} is a_{Fe}. Activity for electrically neutral chemical species is taken as unit. Thus for the above reaction, the Nernst equation will be

$$E = E^0 + (2.303RT/nF) \log_{10}(a_{Fe2+}) \tag{1.10}$$

If $a_{Fe2+} = 1$, then $E = E^0$, the electrode potential at unit ion activity, is defined as 'standard electrode potential,' and is characteristic of the metal. Thus, 'standard electrode potential' is related to ΔG^0, Gibbs energy change in standard state, by the following equation

$$\Delta G^0 = -nFE^0 \tag{1.11}$$

So, it is used to represent the tendency of metal to oxidize. For this purpose, one uses the standard electrode potential series (standard EMF series) (Table 1.3). More negative the potential, higher is the tendency to oxidize. Thus, Table 1.2 shows one extreme of the most passive metal, e.g., Au (with highest or most + value in the EMF series) and another extreme of the most active metal, e.g., K (most negative value of standard EMF).

If we consider a situation where two metal rods one of Cu and another of Zn are immersed in an electrolyte, both metals will develop their redox potential. Let us consider the condition of standard state in equilibrium, the potential of Cu rod is $(E^0)_{Cu2+/Cu} = +0.340$ V (NHE), and that on Zn rod is $(E^0)_{Zn2+/Zn} = -0.763$ V (NHE) (Table 1.3). If both the rods are connected with an electrical wire, there will be net number of electrons flowing from Zn to Cu, thereby stimulating net reduction reaction at Cu rod ($Cu^{2+} + 2e^- \rightarrow Cu$) and net oxidation reaction at Zn rod (Zn $\rightarrow Zn^{2+} + 2e^-$). The consequent path of Zn^{2+} ions will be from Zn to Cu inside

Table 1.3 Standard electrode potential (EMF) series (reduction potentials, potentials are w.r.t. standard hydrogen electrode, at 25 °C)

		Electrode reaction	Standard electrode potential
	Increasingly passive (cathodic)	$Au^{3+} + 3e^- \rightarrow Au$	+1.420
	↑	$O_2 + 4H^+ + 4e^- \rightarrow 2H_2O$	+1.229
		$Ag^+ + e^- \rightarrow Ag$	+0.800
		$Fe^{3+} + 3e^- \rightarrow Fe$	+0.771
		$O_2 + 2H_2O + 4e^- \rightarrow 4OH^-$	+0.401
	↓	$Cu^{2+} + 2e^- \rightarrow Cu$	+0.340
		$2H^+ + 2e^- \rightarrow H_2$	0.000
		$Ni^{2+} + 2e^- \rightarrow Ni$	−0.250
		$Fe^{2+} + 2e^- \rightarrow Fe$	−0.440
	Increasingly active (anodic)	$Cr^{3+} + 3e^- \rightarrow Cr$	−0.744
		$Zn^{2+} + 2e^- \rightarrow Zn$	−0.763
		$Al^{3+} + 3e^- \rightarrow Al$	−1.662
		$Mg^{2+} + 2e^- \rightarrow Mg$	−2.363

the electrolyte, thereby completing the electrical circuit. The current in the circuit will flow in anticlockwise manner. According to the principle of electricity, with this direction of current flow and the sign of the electrical potential of the two metals, Zn rod will be termed 'anode' and 'Cu' cathode. One also observes that, in this condition, Zn electrode will continue to dissolve as Zn^{2+} ions into the electrolyte while Cu electrode will be covered by deposits of various Cu and zinc compounds, generally called as 'corrosion products.' This leads to another important principle of corrosion—**metal with more negative (lesser) potential acts as 'anode' and gets corroded** (dominant reaction at this electrode is *oxidation*), **while metal with less negative (higher) potential acts as 'cathode'** (dominant reaction at this electrode is *reduction*) **where corrosion products form**.

Since corrosion occurs due to a combination of anodic and cathodic reactions, it is always caused by the formation of an electrical cell such as the case discussed above. Thus while discussing various conditions where corrosion may occur, one needs to look at the possibility of formation of electrical cells. In most situations, there is a possibility of three types of electrical cells, namely (i) galvanic cell, (ii) concentration cell, and (iii) oxygen cell.

- Galvanic Cell

This cell forms whenever two different metals are in contact (electrical) with each other. Thus, a steel pipe may be connected to brass valve in a piping system. Galvanized iron (GI) sheet has a coating of Zn over steel, etc. Since the two metals will have different potentials (each metal has different standard electrode potentials), one of the metals acts as anode and other as cathode (similar to the case of Cu-Zn cell described above). Since the two are in electrical contact, current flows from cathode to anode resulting in corrosion of anode as an electrolyte flows in the system (a pipe–valve contact is a part of piping system meant for carrying liquid media, similarly many water systems are made of GI pipes). Metal thus corroding is a case of 'galvanic corrosion.'

- Concentration Cell

Even though the metal in an industrial system may be same (e.g., a storage tank), it may experience corrosion due to variation in concentration of metal ions. According to Nernst equation (Eq. 1.10), a metal having lower concentration of its ions in its vicinity acts as anode as compared to the same metal with higher concentration of metal ions which acts as cathode. Since the two are in electrical contact (being part of the same storage tank), this will result in corrosion of the metal part with lower metal ion concentration in its vicinity.

- Oxygen Cell

Oxygen remains dissolved in water and aqueous solutions. It is responsible for the O_2 reduction reaction Eq. 1.5a in neutral/alkaline media and 1.5b in acidic media. For neutral media, the Nernst equation for oxygen reduction is

Table 1.4 Galvanic series in seawater[a]

Cathodic or passive end	Graphite
	Platinum
	Hastelloy C-276
	Titanium and titanium alloys
	300 series stainless steel
	Monel
	70-3 copper–nickel
	90-10 copper–nickel
	400 series stainless steel
	Silicon bronze
	Admiralty brass
	Copper
	Red brass, yellow brass, naval brass
	Steel
	Cast irons
	Aluminum alloys
	Zinc
Anodic or active end	Magnesium

[a]Relative position of the alloys may change in (i) poorly aerated water and (ii) inside crevices

$$(E)_{O_2/OH^-} = (E^0)_{O_2/OH^-} + (0.059/4) \log_{10}[P_{O_2}/(a_{OH^-})^4] \qquad (1.12)$$

P_{O_2} is fugacity of oxygen which depends upon concentration of oxygen. If P_{O_2} is lower, $(E)_{O_2/OH^-}$ will be lesser and vice versa. So metal area which will be in contact with media having lower oxygen concentration will be 'anodic' and those with higher oxygen concentration will be 'cathodic,' thus forming an electrical cell. In this case, metal part with lower oxygen concentration will corrode. Such a condition is encountered in industry (i) in the presence of crevices (which are quite common) and (ii) in case of partially filled tanks.

- Galvanic Series

Prior to this, we have discussed standard EMF series (Table 1.3). There are two drawbacks while we use it for practical applications: (i) The series mentions only pure metals, whereas, in practice, mostly the materials used are alloys having more than one component. (ii) The placing of a metal in the series is based on its standard electrode potential value. However, in practice, the concentration of metal ions in the solutions does not correspond to unit ion activity. So for industrial applications, one needs to construct another similar series which can tell about the compatibility of two different materials, from the standpoint of galvanic corrosion, when they are used simultaneously as a part of process machinery. Galvanic series (Table 1.4) is such a series. In this series, one considers not only pure metals but also their alloys and the position of a material is based on the equilibrium potential measured at room temperature in seawater, which acts as electrolyte. Although these placings of the material may differ from one media to another, the relative positions do not differ much. However, if need is felt, one can make galvanic series considering materials of interest in the media of interest. Thus, steel and brass, and steel–stainless steel should be avoided unless care is taken against galvanic corrosion of steel, which act as anode of the two metals.

- Kinetics of Corrosion

So far, our discussion is based on thermodynamics of corrosion. However, it is not useful for quantitatively determining the extent of corrosion attack, tendency of metal to experience localized corrosion, passivation behavior, etc. For this purpose, it is necessary to study the changes occurring on the material when its potential is changed from equilibrium or redox value (also later termed as 'open circuit' or 'corrosion potential'). In other words, the material has to be polarized by changing its potential to higher (anodic, e.g., E_a) or lower (cathodic, e.g., E_c) value than the redox value (as defined by Nernst equation, shown in below equation as E_{corr}). If potential is higher than redox potential, the net current will be anodic termed as 'anodic current density,' i_a, representing the oxidation rate of metal. In case when potential is lower than redox value, the current will be cathodic termed as 'cathodic current density,' i_c, representing rate of reduction reactions. These are related to 'polarization' by the following equations:

$$\eta_a = (E_a - E_{corr}) = b_a \log_{10}(i_a/i_{corr}) \qquad (1.13)$$

$$\eta_c = (E_c - E_{corr}) = -b_c \log_{10}(i_c/i_{corr}) \qquad (1.14)$$

b_a and b_c are anodic and cathodic Tafel slopes basically indicating change in potential per decade change of current in the corrosion cell. η_a and η_c are anodic and cathodic polarization, respectively. I_{corr} represents 'corrosion current density' basically representing the rate of oxidation–reduction reaction at corrosion potential. It also represents the rate at which a metal is corroding. To know I_{corr}, b_a, and b_c, a potential versus log(current 'I') curve is recorded. This is known as **'Tafel plot,'** and in recording these plots, the material is polarized within ±250 mV with respect to E_{corr} (or open-circuit potential 'OCP'). Part of the curve above E_{corr} is the 'anodic polarization' curve showing oxidation of metal. Slope of tangent of this curve is b_a. Similarly, part of the curve below E_{corr} is called 'cathodic polarization' curve showing reduction of H^+ to H_2 gas. Slope of tangent to this curve is b_c. From this, one can find out corrosion rate from $I_{corr}(\mu A/cm^2)$ using the following equation:

$$\text{Corrosion Rate } ^{'}R^{'}(\text{mils per year}) = (0.129 \times I_{corr} \times EW)/D \qquad (1.15)$$

where 'EW' is equivalent weight and 'D' density (gm/cm^3). According to this equation, corrosion is decreasing R mils (1 mil = 0.001$''$, 1$''$ = 25.4 mm) thickness of metal each year. From this information, one can know the extent to which thickness of metal sheet will reduce in a given time so as to make it unusable. This helps us in estimating the useful life of the metallic sheet in a given media. E_{corr} indicates the tendency of the metal to corrode. b_a and b_c are used to find corrosion rate using linear polarization curve.

When a material is polarized, in anodic direction, to the extent beyond 250 mV (with respect to E_{corr}), some of these materials, e.g., steels, stainless steels, nickel alloys, etc., show passivation behavior. The recorded polarization curve is referred to as '**anodic polarization curve.**' Thus, one observes that as potential is increased from corrosion potential in anodic direction, current density increases, indicating increased corrosion of metal up to potential Epp (primary passivation potential). In case of passive metals, when applied potential exceeds Epp, corrosion starts decreasing due to the formation of passive layer indicating 'passive' region in curve. On increasing applied potential beyond E_c, the passive film starts breaking which results in onset of pitting and consequential increase in current density. Potential 'E_c' is referred to as 'critical pitting potential.' The lesser is the difference between E_c and E_{corr}, lesser is the resistance of material against pitting. This curve gives information about localized corrosion/passivation characteristics, e.g., (i) whether a material in a given media will experience pitting or not, (ii) range of potential within which material will passivate, (iii) passivation current density if the material is made to passivate, etc.

'**Cyclic polarization curve**' is another polarization curve which tells about the resistance of material against pitting, crevice corrosion, or corrosion in occluded area, e.g., beneath biofilms, etc. According to this curve, after the potential exceeds E_c the material starts pitting until potential E. After potential of metal is increased to E, it is polarized cathodically. Now the current starts decreasing basically indicating that pits are becoming passivated. When potential reaches E_p, the cathodically polarized parts of the curve cut the curve recorded during anodic polarization indicating that all the pits have passivated. The potential E_p is termed as 'repassivation potential.' The lesser the difference between Ep and E_{corr}, lesser is the resistance of material to 'crevice corrosion.'

When material is polarized, cathodically/anodically within ±30–50 mV of E_{corr}, the variation between applied potential and resulting current approximates to linear behavior. This is termed as '**linear polarization.**' Slope of this curve helps one in calculating instantaneous 'corrosion rate' of material and is very useful as an 'online' or 'portable outdoor' corrosion monitoring technique for industries and remotely located pipelines/structures, etc.

Above-discussed polarization curves are 'DC polarization curves' because these are recorded by applying DC potential and measuring resulting direct current (DC). However, useful information on the nature of protective film formed on metal and hence corrosion-resistant behavior of the material can be understood if it is polarized by AC voltage of varying frequencies. The resulting curves obtained are called '**AC**' **impedance curves**. In AC impedance analysis, a corrosion cell is considered as equivalent to an electrical circuit consisting of resistance, inductance, and capacitance and one finds out the impedance of the circuit for different frequencies of AC electrical voltage. A brief idea about the different circuits that can be constructed for a given corrosion process is given now.

The simplest type of corrosion process will be consisting of (i) an oxidation and reduction reaction which proceeds the following activation polarization, e.g., in case of corrosion of steel in 1 M sulfuric acid, and these are represented by Eqs. 1.1 and 1.6; this reaction may be represented by an electrical resistor Rp called 'polarization resistance,' and (ii) the 'double layer' which forms due to + charges on the wall of metal (created due to oxidation reaction) and −ive charges (produced due to oxidation and some negative ions present in the solution itself) diffused into the solution. This double layer is created by the voltage developed at the interface and is equivalent to a 'capacitor.' Solution resistance R_s, of the corroding media, also comes into picture due to a definite nonzero distance between reference electrode (through Luggin capillary) and the metal which is being studied. For this process, the model circuit is shown in Fig. 1.2. Quite often, electrochemical impedance data for such system are affected by 'diffusion' process. The equivalent circuit (Fig. 1.3) for such a situation considers another impedance called 'Warburg impedance' 'W' in series with 'R_p'. Sometimes, a type of equilibrium adsorption of a reaction is intermediately followed by a rapid desorption of the product. This type of inductance is 'pseudo-inductance.' Under such situation, if there is one time constant, the circuit giving rise to the response could be like the one shown in Fig. 1.4. R_p and R_L can be ascertained by comparing the calculated Nyquist and Bode plot with the measured ones. Nyquist plot shows variation of impedance with frequency, while Bode plot is variation between real and imaginary resistances of an impedance circuit.

Fig. 1.2 Circuit for simple impedance response

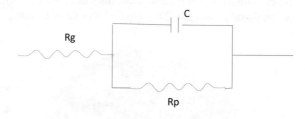

Fig. 1.3 Impedance circuit in the presence of diffusion

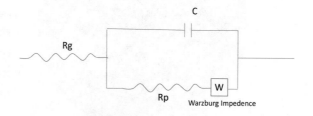

Fig. 1.4 Impedance circuit
in the presence of inductance

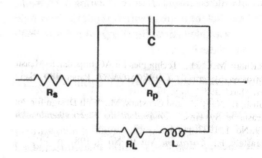

1.2 Types of Corrosion

There are different ways in which corrosion on a material is manifested depending upon the nature of media, composition of material, and the conditions of exposure, e.g., high temperature, high pressure, cyclic stressing, flowing media, etc. Accordingly, one observes eight (8) different types of corrosion. These are (i) general corrosion, (ii) galvanic corrosion, (iii) pitting, (iv) crevice corrosion, (v) intergranular corrosion, (vi) erosion corrosion, (vii) stress corrosion cracking, and (viii) corrosion fatigue. There are some other ways also of naming a type of corrosion, but this one is the most acceptable one. Figure 1.5 shows a rough estimate of % of failure cases due to different types of corrosion in chemical industry (including microbial corrosion also). A brief discussion about the different types of corrosion is now given.

- General (Uniform) Corrosion

When a material's surface is uniformly affected by corrosion, it is called 'general' or 'uniform' corrosion. This type of corrosion results in loss of material thickness

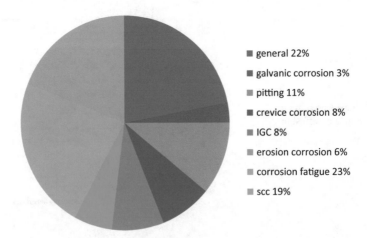

Fig. 1.5 Failure data in chemical industry

Fig. 1.6 Uniform corrosion

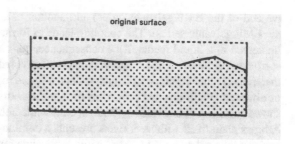

original surface

which reduces its load-bearing capacity (loosely speaking 'strength'). Consequently, the material fails prematurely. This is one of the most commonly observed types of corrosion.

On a polished surface, this type of corrosion is first seen as a general dulling of the surface and, if allowed to continue, the surface becomes rough and possibly frosted in appearance. A polished brass souvenir item turns into a tarnished with dull appearance on passage of time due to corrosion. This type of corrosion appears as uniform thinning of the metal sheet (Fig. 1.6). The reduced thickness may become less than that required for bearing the stress for which the process equipment has been designed. This may result in increased stress for a given applied load. The increased stress may exceed tensile strength of the metal. These sets of events lead to failure of the process equipment. This type of corrosion is of concern where media is acidic, flowing, and/or at high temperature. Often cases of premature failure of hot water heater tube uniform corrosion in ~5 years have been observed. This type of corrosion is monitored by knowing '**corrosion rate**' of the material which defines '*the rate of loss of thickness of metal sheet due to corrosion*' in terms of 'mm per year' or 'mils per year' (1 mil = 1/1000th of an inch). From these data, one can predict expected life of an equipment considering ill effect of corrosion. Failure due to uniform corrosion can be avoided by considering (i) corrosion allowance, (ii) using stainless steel, and (iii) provision of coating/lining/cladding by corrosion-resistant material.

- Galvanic (Bimetallic) Corrosion

This type of corrosion occurs at the coupling of two different metals (Fig. 1.7). This type of corrosion occurs due to one of the metals acting as anode (toward −

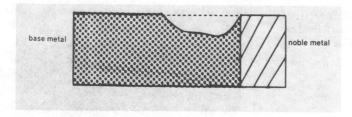

base metal

noble metal

Fig. 1.7 Galvanic corrosion

ive end of the EMF/galvanic series) and another as cathode (toward +ive end of the EMF/galvanic series). The coupling leads to formation of a galvanic cell when immersed in a liquid media. As a consequence, corrosion current flows leading to corrosion of anode, while cathode remains unaffected. Coupling of two different materials is quite common in industry, since a single material is not enough because of different conditions and properties of materials. In Fig. 1.7, schematic diagram shows corrosion of base metal (anode of cell) while noble metal remains unaffected. A brass plate fixed with steel rivets presents a common example where steel rivets being anodic get corroded and fail while brass plate remains as such.

To avoid galvanic corrosion, one of the following steps need to be taken: (i) Ideally speaking, metal should be same as far as possible. Otherwise, metal piece that is to be protected should be noble as compared to the other metal of the couple. (ii) The metals should have as less difference of equilibrium potential as possible, or they should be as close as possible in galvanic series. Less potential difference between the two metals will result in less current, hence less corrosion of the anodic material. Thus in steel–stainless steel, coupling steel will corrode at higher rate than steel corroding in steel–brass coupling, and (iii) provision of insulation to break electrical contact between anode and cathode material. This is equivalent to breaking electrical circuit, hence no current flow so no corrosion of anode material.

- Pitting (Pinhole Attack)

Pitting is one of the localized types of corrosion, in which corrosion affects only some part of the metal whereas the rest remains unaffected. Figure 1.8 shows different shapes of pits formed on a metal surface. Pitting is dangerous because (i) once it initiates, it propagates with ever-increasing rate along the thickness. (ii) It is not apparent as the pitting area is covered with corrosion products. One realizes when liquid starts leaking from the pits. (iii) It is responsible for most number of unexpected/premature failures. Pitting occurs in case of (i) solutions having chemicals, e.g., Cl^-, Cl_2, OCl^-, ClO_2, $S_2O_3^{2-}$, S^{2-}, etc., and (ii) stagnant solutions which enhance the concentration of chemicals. It is severe at low pH and high temperature. Pitting generally occurs on those materials which form a protective layer, e.g., stainless steels, nickel alloys, etc. When protective layer fails at some places, the smaller area of fresh metals, exposed to corrosive media, acts as anode and covered area, which is larger, acts as cathode. This gives a cell having large cathodic and small anodic area which is responsible

Fig. 1.8 Different sizes and shapes of pits formed on metal surface

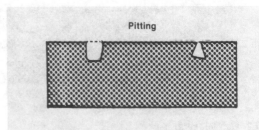

for severe localized attack at anode. Pitting can be avoided by using pitting-resistant materials, by avoiding pitting-causing chemicals or by anodic/cathodic protection techniques by utilizing passivation behavior of materials, e.g., stainless steels. Due to formation of biofilms, in most cases of microbial corrosion, the metal shows pitting attack.

- Crevice Corrosion (Corrosion under Deposit)

This type of corrosion occurs at a place where liquid media may get trapped. The trap may be large enough for entry of the liquid media but small enough so that liquid once entered is not able to move out or get trapped. Figure 1.9 shows the mechanism of crevice corrosion.

Examples of crevice corrosion are (i) corrosion of flange metal by crevice formation because rubber gasket results in crevice corrosion of flange in piping system. (ii) Crevice corrosion also occurs due to some faulty design, which if taken care may avoid this type of crevice corrosion. (iii) In case of joining of two metal plates by a rivet/screw, the area between rivet and plate surface or between the two plates acts as crevice which corrodes much more than rest of metal surface. Accordingly, joints between metal parts, in industry, are always suggested to be done by welding, wherever possible.

- Intergranular Corrosion (IGC)/Weld Decay

This type of corrosion occurs due to change in the chemical composition of some engineering materials along their grain boundaries such that grain boundary area becomes more prone to corrosion than the rest of the material. This change usually occurs due to exposure of material to high temperature >400 °C, for a long duration. This could be the case of (i) material being part of machinery meant for high-temperature processes, (ii) material having undergone welding operation for joining with other material, hence called as 'weld decay,' and (iii) due to 'stress relieving.' The corrosion attack is generally observed in the heat-affected zone (HAZ) (Fig. 1.10).

Stainless steels or Cr-containing alloys are a good example of this type of corrosion. In such cases, heating to high temperature results in Cr changing to Cr_2O_3. This change is usually found in or around grain boundary. Thus, grain boundary becomes depleted in Cr which is responsible for providing corrosion resistance to stainless steel. As such, the steel, in the heat-affected zone, experiences corrosion.

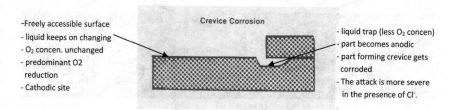

Fig. 1.9 Mechanism of crevice corrosion

Fig. 1.10 Intergranular corrosion (weld decay)

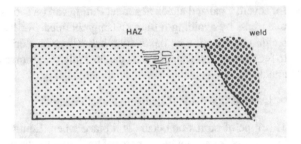

- Erosion Corrosion

This type of corrosion occurs due to combined action of mechanical stress, from motion of fluid/liquid, and corrosive action of chemicals present in the liquid media. Schematic diagram of this type of corrosion is shown in Fig. 1.11. The wearing action, due to flow of liquid, removes the protective film causing fresh metal surface to expose. The fresh metal experiences higher corrosion attack, and again a protective film is formed. Thus, reformed protective layer is again removed by the erosive action. Again fresh surface of metal becomes exposed to corrosive liquor. Overall, the recursive action of formation and removal of protective layer (layer of corrosion products) enhance corrosion significantly. Examples of erosion corrosion are observed in pump impeller and pipe system at bends or at low cross-sectional areas due to increased flow velocity. Erosion corrosion is promoted by high flow rate, less hardness of material, turbulent conditions, suspended particles in fluid.

- Stress Corrosion Cracking (SCC)

This type of corrosion occurs when stress and corrosive environment are simultaneously present.

The attack appears in the form of cracks on the material surface (Fig. 1.12).

Many times, these cracks initiate and propagate inside the material (insidious type of attack); hence, they do not show on surface and they keep on propagating until the material fails unexpectedly. Not all chemical–metal combinations are susceptible to SCC. Chemical–metal combinations responsible for SCC on metal are shown in

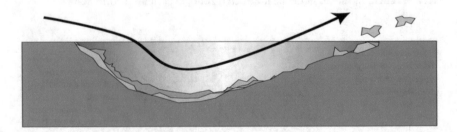

Fig. 1.11 Erosion corrosion

Fig. 1.12 Stress corrosion cracking

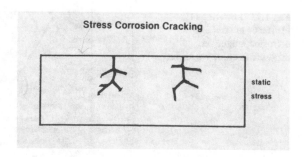

Table 1.5 Material–environment combination susceptible to stress corrosion cracking

Material	Chemical environment
Carbon steel	Caustic, sulfides, nitrates, hydrogen
Austenitic stainless steel	Chloride, acidic hydrogen sulfide, hydroxide
Cu alloys	Ammonical based solutions
Al alloys	Seawater (chlorides)

Table 1.5. Temperature, stress level, and chemical composition of alloy also play an important role in inducing stress corrosion cracking.

- Corrosion Fatigue

This type of attack occurs in the presence of cyclic stress and corrosive media. It leads to failure of material due to fatigue which occurs at much lower stress than yield stress in the presence of corrosive media.

Application of a material in case of cyclic stress conditions is determined by the reduction of its strength as it experiences increasing number of stress cycles. One defines 'fatigue limit' or 'endurance limit,' a parameter which represents the maximum stress that can be applied on material so that it does not fail by fatigue or maximum stress at which material fails after experiencing 10^8 cycles. The attack is manifested in the form of unbranched cracks which weakens the material. Fatigue limit of a material depends on its environment. Figure 1.13 shows lowering of fatigue limit of steel, with respect to that measured in air, as it is exposed to water and sodium chloride solution. Thus, steel is expected to fail at a lower stress level (~200 MPa) in NaCl solution as compared to air when it can survive up to ~300 MPa stress level. The fatigue fracture is brittle, and the cracks are most often transgranular, as in stress corrosion cracking, but not branched. The corrosive environment can cause faster crack growth and/or crack growth at a lower tension level than in dry air. Even relatively mild corrosive atmospheres can reduce the fatigue strength of aluminum structures, considerably, down to 75–25% of the fatigue strength in dry air. This type of corrosion is observed in case of (i) press rolls in paper machine, (ii) material of construction of vibrating structures, e.g., stranded cables, (iii) tubes in steam boilers where tubes are subjected to thermal cycling and resulting thermal stress cycles,

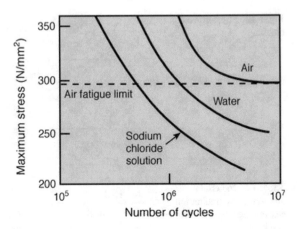

Fig. 1.13 Lowering of resistance of steel against corrosion fatigue on exposure to corrosive media

and (iv) in oil industry, exposure of drill pipe and sucker rods to brines and sour crude. To minimize case of failure due to corrosion fatigue, newer materials are being developed. These materials are stronger and stiffer and have better resistance against corrosion fatigue. These newer alloys are Duplex SS VKA-78, Martenitic SS CA-15, Tin Bronze, etc.

1.3   Corrosion Measurement and Testing

Now an account of various methods, used in estimation of the extent of corrosion, is given in this section. Two types of laboratory tests are conducted for this purpose— long-term test and short-term test. Both tests have equal importance and in fact compliment each other's finding. Long-term tests are usually carried for a duration covering several weeks/months due to (i) need of knowing about the performance of material (as a part of industrial machinery and infrastructure) against corrosion including microbial corrosion in order to know their suitability to survive for couple of years, and (ii) localized corrosion, e.g., pitting, crevice corrosion, etc., may take few weeks before they initiate and show their deteriorating effect on the test material. Short-terms tests may consume few hours only to know about the suitability of candidate materials, so that they do not fail prematurely in an aggressive environment. Since there can be several types of materials suitable for a given corrosive media, one can perform short-term test on them to short-list maybe 3–4 of them and then perform long-term test on these short-listed materials to check their appropriateness for their possible usage as industrial machinery/infrastructure for ~10 years or so without any maintenance. The short-term tests also help in designing suitable protection system against corrosion for a material. First, we describe about the laboratory methods. Those methods which are used for in-plant test or online monitoring have been dealt with in the next section.

- Immersion Test

In these tests, the test material is kept immersed in the liquid media in which corrosion resistance of the material is to be checked. The immersion time may vary from few days to several weeks or sometimes even several months depending upon the corrosivity of the media. If one expects localized corrosion also to be observed, the immersion time may be kept several months because sometimes it takes quite long time for localized corrosion to initiate. For these tests, the material is taken in the form of a plate so that it has a high surface area. Prior to immersing the material in the test media, the material is cleaned and polished to remove dirt, oil, scratches, etc. An acceptable test material should have a mirror-finished surface. One then measures its area and its weight. The sample is also cleaned ultrasonically to remove any dirt particles from microsized cavities, if they are there. Finally, the specimen is degreased by acetone/methanol to remove any oil/grease which may appear on metal surface by touching with hands. Thus, cleaned and weighed sample is put for exposure by immersing in the test media. The amount of test solution is taken on the basis of ~250 ml/square inch of exposed surface area. The duration of tests is decided by the following equation:

$$\text{Time (in Hours)} = 2000/\text{Corrosion rate (mils per year)}$$

The corrosion rate taken should be an estimated value. In addition to information on the extent of uniform corrosion, information about pitting, crevice corrosion, and weld-related attack can also be obtained. For this purpose, the test sample should be welded (Fig. 1.14) and should be kept in the test media with serrated washers with an arrangement as shown in Fig. 1.15.

After the sample has been exposed for a predetermined duration, it is taken out and cleaned to wash off corrosion products. The washers are removed, and the corroded sample is weighed. The corroded sample might appear with excessive attack near welded part (Fig. 1.16).

The extent of uniform corrosion is calculated by the following formula:

$$\text{Corrosion Rate (mils per year)} = 534 \times W/(D \times A \times T) \qquad (1.16)$$

Fig. 1.14 Welded material coupon for immersion test

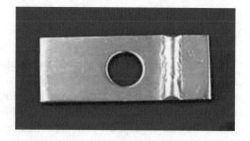

Fig. 1.15 Coupon with serrated washer and crevice corrosion on the coupon

Fig. 1.16 Excessive corrosion on welded portion of stainless steel 316L coupon

where W is weight loss of coupon (in mg), D—density of material of coupon (in g/cm^3), A—area of coupon (in inch^2), and T—time of exposure (in hours). Corrosion rate indicates the average loss of thickness of metal coupon (in mil—10^{-3} inch) per year. This value can help us to estimate useful service life of a given material in a given environment. It also helps us to calculate 'corrosion allowance' (additional thickness of metal sheet) so as to ensure a service life of fixed number of years, e.g., 10 years/15 years, etc. One can then look under the microscope to evaluate the depth of pits formed on open surface (pitting), under serrations of serrated washers (crevice corrosion), and near the welded area (Fig. 1.16) for estimating weld-related attack.

Test for estimating performance of material against stress corrosion cracking is performed by putting samples in stressed form in corrosive media. After the test, metal coupons are evaluated for extent of corrosion attack through weight loss and appearance of cracks on the strained surface.

- In-plant Test

These tests are carried on several candidate materials, for constructing plant machinery, simultaneously by exposing them in some section of a plant, e.g., boiler, pipeline, pressure vessel, etc. Accordingly, the result of these tests shows the performance of material against corrosion in real-life environment. This information helps in selecting the appropriate material meant for constructing the part of plant in which the test has been performed. First, the metal samples are cleaned as described in section on immersion test, then they are weighed, and their surface area measured.

Fig. 1.17 Rack for in-plant test

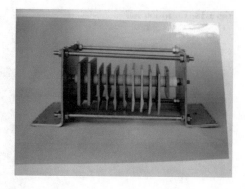

The metal samples, for the test, are fixed in rack (Fig. 1.17) with serrated washers to avoid electrical contact between two successive metal samples. This way one avoids the possibility of the metal samples experiencing galvanic corrosion.

The metal samples should be tightened enough so that they do not move during the test. This is done to avoid samples experiencing erosion corrosion. The rack is then fixed by welding/screwing/riveting inside or on a machinery part for which we intend to select the material of fabrication. This place could be inside pipeline, inside boiler/pressure vessel/storage tank or on washer, etc. The rack is kept for exposure for a duration of ~6 months so that one can have an idea about the long-term performance of test metals in the service environment. After the exposure period is over, the rack is removed and corroded samples are taken for cleaning. Afterward, the corroded samples are weighed to find weight loss and their corrosion rate. The samples are also viewed under microscope to measure the depths of pits formed on open surface (pitting), under crevice or deposits (crevice corrosion) (Fig. 1.15) or near welded part (weld-related attack) (Fig. 1.16). This way one assesses about the resistance of metal against uniform corrosion, pitting, crevice corrosion, and weld-related attack.

- Electrochemical Test

There are several types of tests performed in this category. These are: (i) determination of open-circuit potential, (ii) Tafel plot determination, (iii) anodic polarization and cyclic polarization test, (iv) potentiostatic test, etc. Before a description of these tests is given, a brief account of experimental setup is given for performing these tests.

One of the main parts of this setup is corrosion cell (Fig. 1.18), which is a five-/six-neck cell. The sample to be tested is put in this cell as working electrode in an arrangement shown in Fig. 1.19. The sample considered here is a piece of rod (cylindrical); however, holders for flat specimens are also available commercially. There are two counter/auxiliary electrodes for maintaining flow of current and a reference electrode for monitoring potential of working electrode (test sample). As can be seen, the cell has inlets/outlets for gas and for thermometer also. This cell forms part of the electrical circuit consisting of a potentiostat/galvanostat and a controlling/recording device in the form of a computer. With the help of an appropriate software meant for various electrochemical measurements for corrosion testing, the

Fig. 1.18 Corrosion cell

Fig. 1.19 Cylindrical
sample in electrode holder

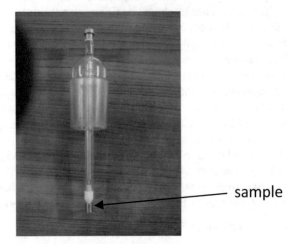

potential of the working electrode is changed with a given scan rate and the resulting
current is recorded and displayed on computer monitor. The software helps in calculating various parameters related to corrosion performance (discussed earlier) of the
test material (working electrode). These tests have been described in the following
lines.

E versus time—This test is meant for knowing open-circuit potential. Sample is
polished to avoid cracks and cleaned with acetone to avoid dust and grease. The
sample, in cylindrical/plate form, is fitted in working electrode holder of corrosion
cell (Fig. 1.19). The sample holder is immersed in corrosion cell, and potential is
measured at fixed interval of time for a duration such that slope of E versus time
curve is almost zero. Total time for test may vary from few hours to few days even.
The saturated value of potential indicates open-circuit potential (OCP).

Linear Polarization Resistance Measurement: Sample, in electrode holder, is immersed in test solution with reference electrode (saturated calomel electrode) and counter electrode connected through a corrosion measurement system. The sample is allowed in this state for about an hour for equilibrium to achieve. Afterward, the sample is polarized within $E_{corr} \pm 10 - 25$ mV from cathodic to anodic direction at a scan rate of 0.6 V/h or 0.166 mV/s. A linear behavior between E and resulting current density is observed. The slope of E versus i curve is measured to know R_p and i_{corr} using equation

$$[d(\Delta E)/di]_{\Delta E=0} = R_p = \beta_a\beta_c/2.303(\beta_a + \beta_c)i_{corr} \qquad (1.17)$$

This is followed by calculation of corrosion rate using equation

$$\text{Corrosion rate (mils per year)} = 0.129 \times I_{corr} \times \text{Eq.weight}/D$$

where D is density of test metal (gm/cm^3), I_{corr} in $\mu A/cm^2$.

Tafel Plot—Sample is immersed and kept in liquid media as described earlier. Afterward, it is polarized within $E_{corr} \pm 250$ mV from cathodic to anodic direction. Tangents drawn to anodic and cathodic part should meet at E_{corr} and i_{corr}. Slope of anodic part is β_a and that of cathodic part is β_c. Tafel slopes also indicate as to which of the anodic/cathodic polarization is dominating.

Anodic Polarization Curve—For recording this curve, the sample is polarized in anodic direction between $+800 \geq 1200$ mV w.r.t. E_{corr} with scan rate of 0.6 V/h. This curve gives information about passivation behavior of metal. Parameters, related to passivation, are obtained from this curve, e.g.,

E_c—critical pitting potential, E_p—repassivation potential, I_{crit}—critical anodic current density for passivation, I_{pass}—passivation current, passivation range ($=E_c - E_{corr}$) or margin of safety against pitting.

A comparison between E_c, E_{corr}, and E_p indicates about the following aspect of localized corrosion on the given material

$E_{corr} > E_p$	Material will experience pitting and crevice corrosion
$E_c > E_{corr} > E_p$	Material will experience crevice corrosion but not pitting
$E_{corr} < E_p$	Material is resistant to pitting as well as crevice corrosion

Addition of chemicals resulting in increase of E_{corr} or decreasing E_c and/or E_p shows increase of tendency of material to experience localized corrosion. In other words, addition of such chemicals is likely to induce localized corrosion on test metal. Other tests, e.g., linear polarization, cyclic polarization, potentiostatic, etc., are also performed on materials in various electrolytic media. Their details can be obtained from any book dealing with corrosion basics [4].

- Monitoring of Corrosion (Online/In-service Testing)

These tests provide information on corrosion rates or corrosion attack during the plant operational condition. It, therefore, monitors the condition of plant 24×7 against corrosion as it operates and may alarm about its impending failure if material of construction of plant has reached the limit of failure. It also indicates about real-time changes in corrosivity due to change in process conditions as a result of addition of new chemicals or change in temperature, pH, etc. The following techniques are adopted for monitoring online corrosion in plants: (i) electrical resistance method, (ii) linear polarization resistance (LPR) method, and (iii) ultrasonics.

Electrical Resistance Method—Instead of mass loss, in this method, one measures the electrical resistance of wire or a suitable device. Corrosion decreases the thickness of flat specimens or diameter of wire. This leads to increases in electrical resistance 'R' of the probe material with time 't'. The slope of R versus t curve gives corrosion rate. The advantage of this technique is (i) small probes—easy installation, (ii) continuous monitoring with PC, and (iii) most environments do not affect the conductivity except molten salts. However, the disadvantages are (i) potential leaks due to probe penetration, (ii) need of temperature compensation device since electrical resistance is affected by temperature also, and (iv) this technique gives no information on localized attacks, e.g., pitting, crevice corrosion, stress corrosion cracking.

Linear Polarization Resistance (LPR) Method—This method estimates corrosion current using Stern–Geary equation

$$I_{corr} = \left(1/R_p\right)\beta_a\beta_c/(2.303(\beta_a + \beta_c)) \tag{1.18}$$

R_p is measured from linear polarization curve. Corrosion rate is calculated from the following equation

$$\text{Corrosion Rate } 'R'(\text{mils per year}) = (0.129 \times I_{corr} \times \text{EW})/D$$

In this method, probe of the material of interest is placed in plant. It is attached with system which polarizes the probe within ± 10 mV with respect to open-circuit potential of the probe material and calculates R_p and corrosion rate. It alarms as and when very high corrosion rates are reached during operation of the plant.

Ultrasonic Thickness Measurement—This technique uses ultrasonic wave for measuring thickness of metal sheet. The sheet could be a wall of tank, reaction vessel or pipe line, etc. In this technique, an ultrasound transducer is fitted against a vessel exterior. The ultrasound travels through the metal and returns back from the other end. Since ultrasound travels only in a medium, it returns from the point wherever there is a discontinuity of the medium. Thus, it also helps in detecting cracks present inside the metal. The thickness of the metal is calculated as given below:

$$\text{Thickness} = 1/2(\text{time taken by ultrasound for return journey from transducer}$$
$$\times \text{ velocity of ultrasound})$$

Ultrasonic technique can be used for diversified applications, e.g., for inspecting the presence of cracks inside the railway track to check its vulnerability to fail due to weight of the train. In the same way, it can also be used to determine residual thickness of wall of a tank to determine the vulnerability of tank to failure due to corrosion.

References

1. Bhaskaran R, Palaniswamy N, Rengaswamy NS, Jaychandran M (2005) A review of differing approaches used to estimate the cost of corrosion (and their relevance in the development of modern corrosion prevention and control strategies). Anti-Corrosion Methods and Materials 52(Iss1):29–41
2. Koch GH, Brongers Michiel PH, Thomson NG, Viramani YP, Payer JH (2002) Corrosion costs and preventive strategies in the United States, NACE, Publication No. FHWA-RD-01-156
3. Beech IB, Gaylarde C (1999) Recent advances in the study of biocorrosion—an overview. Revista de Microbiologia 30:177–190
4. Singh AK (1995) Corrosion and its control in paper industry. AICTE, New Delhi

Chapter 2
Microbial Induced Corrosion and Related Theories

Abstract After introducing corrosion, the next step obviously is to learn about Microbial Induced Corrosion (MIC). Fundamentals of MIC are same as those for corrosion. But a basic difference between the two is in terms of significant role played by microorganisms in corroding the materials. By way of various metabolic reactions and deposition of biofilms, the microbes change the chemistry of the environment which help in inducing corrosion or accelerate corrosion reaction rates. However, as you shall see, while going through the later chapters of the book, that microbes not only induce corrosion but some inhibit corrosion. This chapter thereafter discusses various aspects related to MIC such as biofilm formation, bioenergetics, on the basis of electrochemical fundamentals, leading to initiation of MIC, Biocatalytic Reduction, Extra cellular electron transfer, carbon source starvation etc. The chapter then discusses MIC due to some important bacteria e.g. sulfate reducing bacteria (SRB), Iron oxidizing bacteria (IOB), Nitrate reducing bacteria (NRB), Fungi, Algae, Archaea.

Keywords Biofilm · Metabolic reaction · Bioenergetics · Biocatalytic reduction · Extracellular electron transfer

Research investigations as well as industrial studies including in-plant tests related to chemical corrosion have been known since many decades, but microbial induced corrosion (MIC) is a relatively new phenomenon in the field of material deterioration. It is also referred to as 'biocorrosion.' As the name implies this phenomenon relates to corrosion of material induced by the presence of microbes. A non-corrosive biofilm usually means that it cannot cause corrosion by itself. However, when CO_2 or another pre-existing (often non-biogenic) corrosive agent is abundant, a non-corrosive biofilm may damage or destroy the $FeCO_3$ protective film on a steel surface during its metabolism. This can accelerate CO_2 corrosion greatly. This kind of induced corrosion still belongs to MIC. Work related to this area is also currently being pursued on the inhibition effect of microbes to control microbial induced corrosion in the presence of some other bacteria. So this field is likely to be redefined soon as 'microbial induced corrosion and inhibition' (MICI). Work done until now has more or less established that MIC is also a type of electrochemical corrosion where

© The Author(s), under exclusive license to Springer Nature Singapore Pte Ltd. 2020
A. K. Singh, *Microbially Induced Corrosion and its Mitigation*,
SpringerBriefs in Materials, https://doi.org/10.1007/978-981-15-8019-2_2

Table 2.1 Bacteria responsible for influencing corrosion

S. No.	Bacteria
1	SRB
2	Metal-oxidizing bacteria
3	Metal-reducing bacteria
4	Metal-depositing bacteria
5	Slime-producing bacteria
6	Acid-producing bacteria
7	Fungi
8	Pseudomonas

various electrochemical reactions (oxidation and reduction reactions) are responsible for corrosion/inhibition of metals, and the role of bacteria is in one way or the other is to change chemical conditions such as to produce either more corrosive chemicals so as to enhance corrosion or bring changes, e.g., reduce the dissolved oxygen concentration, forming a protective biofilm, etc., so as to inhibit corrosion. Thus, a particular liquid media, in the absence of corrosion causing bacteria, may not attack a metal significantly. However, in the presence of bacteria, the metabolic reactions may result in change in the chemical environment such as to enhance metal corrosion significantly or inhibit MIC. Prior to discussing theories related to MIC in the presence of various microbes, it will be important to know about the microbes responsible for influencing corrosion. Table 2.1 gives a list of some such bacteria:

2.1 Biofilm Formation

In general, whenever a metal surface gets exposed to a liquid media infested with bacteria, they start attaching with the metal surface. The phenomenon of bacterial adhesion is an important phenomenon for those working with microbial corrosion. This is so because the corrosion effect by bacteria on a given material is induced by the presence of bacteria in close proximity of the surface of that material.

Once the bacteria get adhered to the metal surface, more and more bacteria become attached to it, thereby covering the metal surface with what is called 'biofilm.' Biofilm commonly is observed to be responsible for bacterial growth and, therefore, has an important impact on corrosion of materials.

Biofilm formation is understood to consists of five stages: (1) initial attachment, (2) irreversible attachment, (3) maturation I, (4) maturation II, and (5) dispersion. These are described below:

Stage 1 is when the bacteria are free-floating type (planktonic stage) within an aqueous environment. Bacteria, however, prefer to grow on available surfaces (sessile form) rather than existing in planktonic form. Hence, on reaching surface, they get attached reversibly through weak Van der Waals bonds or electrostatic (due to surface being positive charged and bacteria negatively charged). At this stage, they can be removed from surface by scrubbing action. Stage 2 is 'irreversible attachment,' where the bacterium first gets adsorbed on the surface, and then the cell actively attaches itself to the surface by its metabolic activity. This is the stage where extracellular polymeric substance (EPS) form which are responsible for irreversible attachment of bacteria. Irreversible bacterial adhesion occurs because of short-range molecular interactions like hydrogen, ionic, and covalent bonding, interactions involving extracellular structures, and secretions. The cell secretes a substance: A polysaccharide known as glycocalyx (hydrated polymeric slimy matrices), which subsequently enables the bacterium to encapsulate itself on the surface and leads to biofilm formation. Otherwise, with time, their adherence to the surface become stronger and permanent through cell adhesion structure leading to formation of microcolonies. These microcolonies grow with time resulting in growth of the biofilm. It is important to note that not all bacteria produce slime and that bacterial strains that do not produce slime are less adherent. Such bacteria may not induce corrosion on metals. In stage 3, maturation I and stage 4, maturation II, with further passage of time, biofilm formed in previous stage provides bacteria with an environment that is highly conducive to their survival while also protecting them from various types of aggression from outside, such as flows of liquids and changes in pH or temperature. Consequently, overtime, a majority of microorganisms become associated with them. Thus, large sessile bacterial biofilm populations always outnumber the planktonic (floating) populations of the same systems by a factor between 1000 and 10,000. In 5th stage—the last stage, the dispersal of cells from the biofilm colony is an essential stage of the biofilm life cycle. Dispersal enables biofilms to spread and colonize new surfaces. Within the mature biofilm, there is a bustling community that actively exchanges and shares products that play a pivotal role in maintaining biofilm architecture and providing a favorable living environment for the resident bacteria. However, as biofilms mature, dispersal becomes an option. Besides passive dispersal, brought about by shear stresses, other causes for dispersal are alterations in nutrient availability, oxygen fluctuations, and increase of toxic products, e.g., enzymes that degrade the biofilm extracellular matrix, a fatty acid messenger, *cis*-2-decenoic acid, secreted by *Pseudomonas aeruginosa*, etc., or other stress-inducing conditions. Voids within the biofilm are also created by cell death, serving as an additional dispersal mechanism that frees resident live bacteria, as shown by studies in *P. aeruginosa*. Dispersing bacteria have the capacity to reinitiate the process of biofilm formation, on encountering a suitable environment. It is generally assumed that cells dispersed from biofilms immediately go into the planktonic growth phase. However, recent studies have shown that the physiology of dispersed cells from *Pseudomonas aeruginosa* biofilms is different from those of planktonic and biofilm cells. Hence, the dispersal process is a unique stage during the transition from biofilm to planktonic lifestyle in bacteria.

2.2 Initiation of MIC (Bioenergetics)

To initiate a corrosion reaction, there should be an anode which oxidizes and release e^-'s. Metals, which form the engineering materials, e.g., steels, brass, bronze, Al alloys, etc., represent such category of materials. This oxidation also leads to deterioration of anode materials, which results in microbial induced corrosion, in the presence of bacteria or other microbes, of metals. To maintain charge neutrality (conservation of charge) of overall system (a compulsion for corrosion to occur), some reduction reactions should also occur at another site termed as cathode. One also requires some medium which is conducive for flow of e^-'s, through the liquid medium, from anode to cathode. This possibility occurs in the presence of bacteria, in addition to that due to the ions present in liquid media, in the medium through the development of biofilm over metal. Earlier, a clear explanation of bioelectrochemical processes occurring at the interface of the biofilm and metal surface was not in sight. Now, MIC can be understood on the basis of bioenergetics and extracellular electron transfer (EET) [1–5].

In a natural environment, microorganisms normally exist in a biofilm formed of extracellular polymeric substances (EPS) consisting mainly of proteins, carbohydrates, lipids, etc. This is possible due to availability of energy, for bacterial growth, through electrochemical reactions (1.5b and 2.1). In an anaerobic media, in case of MIC due to sulfate-reducing bacteria (SRB), the reduction reaction could be due to sulfate as a constituent of the growth media, leading to formation of sulfide

$$SO_4^{2-} + 8H^+ + 8e^- \rightarrow S^{2-} + 4H_2O \quad \text{(Reduction)} \quad E^0 = -0.217\,\text{V} \quad (2.1)$$

$(\Delta G^0)_{\text{reduction}} = nFE^0 = 8 \times 96500 \times (-0.217)\,\text{J}/(4.184\,\text{Jcal}^{-1}) = -40.04\,\text{kcal/mole}$ of sulfide. ΔG^0 and E^0 are Gibb's free energy in standard state and standard electrode potential, respectively.

Oxidation of lactate (constituent of media) to acetate and CO_2 (electrochemical data [5])

$$2 \times [CH_3CHOHCOO^- + H_2O \rightarrow CH_3COO^- + CO_2 + 4H^+ + 4e^-] \text{ (Oxidation)}$$
$$E^0 = -0.430\,\text{V} \quad (2.2)$$

$(\Delta G^0)\text{oxidation} = nFE^0 = 8 \times 96500 \times (-0.430)\,\text{J}/\left(4.184\,\text{J cal}^{-1}\right) = -79.32\,\text{kcal/mole}$

Acetate and lactate are among the hydrocarbons which act as a source of organic carbon for bacterial growth. Available energy as a consequence of the above two reactions for bacterial growth is

$(\Delta G^0)\text{oxidation} - (\Delta G^0)\text{reduction} = -79.32 - (-40.04) = -39.28\,\text{kcal/mole of sulfide}$

Vulnerability of a metal to corrosion depends on whether it can be used as e^- donor for microbial metabolism which provides energy required for bacterial survival in the media. Let us consider that, in the presence of bacteria, corrosion of metal proceeds by following an additional oxidation (Eq. 1.1). One has to consider production of $8e^-$'s due to this reaction by considering oxidation of 4 iron atoms, so that these electrons reduce 1 sulfate ion to sulfide (Eq. 2.1) so that

$$(\Delta G^0)_{oxid} = nFE^0 = 8 \times 96500 \times (-0.447) \, \text{J}/\left(4.184 \, \text{Jcal}^{-1}\right) = -82.47 \, \text{Kcal/mole}$$

Thus, 1 mol of sulfide is produced per 4 mol of Fe^{2+}. In the whole system, electrons produced from oxidation (Eqs. 1.1 and 2.2) are consumed by reduction (Eq. 2.1) in the presence of SRB. The potential difference in electrochemical cell due to reduction (cathode) and oxidation (anode) considering, respectively, Eqs. 1.1 and 2.1 are greater than that when one considers oxidation due to Eq. 2.2. Consequently, driving force for iron oxidation will be higher than that for oxidation of lactate. Thus, in the presence of iron as steel in the inoculated media, energy available for survival of bacteria $= -82.47 + 40.04 = -42.43$ kcal/mole. Consequently, iron continues to corrode microbially while bacteria survive due to availability of energy.

The reduction of oxidant, e.g., oxygen, sulfate, etc., in above cases occur inside a cell's cytoplasm because it requires biocatalysis by intracellular enzymes. When one considers only bacteria in media, both oxidation of lactase and reduction of oxygen/sulfate takes place inside cell itself because all of them are present there. However, when it comes to MIC of metal in a microbial media, e^-s source is oxidation of metal which is outside the cell because metal is not soluble. Hence, one has to consider some way of transferring e^-s to inside the cell so that reduction may take place due to presence of biocatalyst there. Such transfer of electrons across cell wall is known as extracellular electron transfer (EET). Thus, this EET-MIC is responsible for providing energy for the cell growth and is accompanied by MIC of metal.

- Biocatalytic Reduction: (BCSR/BCNR Theory)

Microorganisms require two components to provide energy for their metabolism (i) electron donor (energy source) which is some metal, e.g., steel, etc., in case of MIC and/or hydrocarbons and fatty acids (formate, acetate, lactate, etc.) which normally provide energy and organic carbon for growth of bacteria and (ii) an electron acceptor. Thus, SRB use sulfate as electron accepter and convert into sulfide on reduction (Eq. 2.1). Additionally, some SRB also utilize thiosulfate, sulfite, etc., also as electron acceptor. Similarly, in some cases, nitrate and nitrite may be used as electron acceptor which undergo the following reduction reaction

$$NO_3^- + 10H^+ + 8e^- \rightarrow NH_4^+ + 3H_2O \quad E^0 = +0.360 \, \text{V} \qquad (2.3)$$

$$2NO_3^- + 12H^+ + 10e^- \rightarrow N_2 + 6H_2O \quad E^0 = +0.760 \, \text{V} \qquad (2.4)$$

$\Delta G^0 = -621$ kJ/mole of nitrate considering redox reactions Eqs. 1.1 and 2.3

$\Delta G^0 = -577$ kJ/mole of nitrate considering redox reactions Eqs. 1.1 and 2.4

In microbial corrosion of steel, thus, the oxidation of lactate and iron and reduction of sulfate or nitrite, etc., form the anodic and cathodic reactions, respectively, of the electrochemical cell. A comparison of their redox potential and change in Gibbs' free energy indicates that the reactions are thermodynamically favorable which produces energy for growth of bacteria. Thus, in MIC, instead of a physical cathode, here, sulfate, thiosulfate, nitrate, etc., act as cathode named as 'biocathode.' Additionally, it is observed that without biocatalysis of bacteria, the reduction reaction has a negligible rate which does not explain aggressive corrosion attack observed by metals due to bacteria. However, experiments have shown that, in the presence of bacterial biofilm, the rate of cathodic reaction is very high due to biocatalysis. This led to suggestion of biocatalytic sulfate reduction (BCSR) theory in case of SBR and BCNR (biocatalytic nitrate reduction) theory in case of nitrate-reducing bacteria *P. aeruginosa* [1]. The bioenergetics of NRB MIC against carbon steel is similar to SRB MIC.

2.3   Extracellular Electron Transfer (EET)

The transfer of electrons (produced by the oxidation of metal) from the vicinity of corroding metal to inside the cell cytoplasm takes place through (i) direct electron transfer (DET), (ii) conductive pilus, and (iii) mediated electron transfer (MET).

In 'DET,' sessile cell walls of microorganisms have a direct contact with the metal surface through cytochrome 'c' or the cells take help of conductive pili (nanowires) to establish contact with a metal surface for electron transfer. In *mediated electron transfer* 'MET,' electron transfer mediators (electron shuttles), dissolved in media, absorb electrons from oxidizing metal's surface and shuttle to release them on the cytochrome 'c' bound to cell wall (Fig. 2.1). Recent studies [6–8] have shown increase in the rate of MIC and severe pitting attack when electron transfer mediators, e.g., riboflavin and flavin adenine dinucleotide (FAD) are added in case of MIC by *Desulfovibrio vulgaris* and *Pseudomonas aeruginosa* on carbon steel and stainless steels 304.

To test the effect of electron mediator, 1018 carbon steel was tested in ATC 1249 medium, in abiotic state and later inoculated with SRB *D. Vulgaris*. Corrosion experiments were carried in the test media with and without addition of electron mediators FAD and riboflavin. After exposure of 7 days, the average weight loss was observed 0.2 mg/cm^2 for abiotic media. No significant change in weight loss was observed on addition of mediators suggesting that the mediators were not corrosive. However, in the presence of SRB in the media, average weight loss was observed 2.1 mg/cm^2. When electron mediators were added in the SRB inoculated medium, corrosion rate was observed to increase to 3.4 and 3.1 mg/cm^2 when 10 ppm FAD

Fig. 2.1 Extracellular
electron transfer methods
applicable in MIC from
metal to inside the cell for
metal corrosion and cell
growth

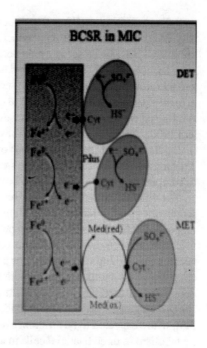

and riboflavin were added, respectively, indicating an increase of 62 and 48% in weight loss over the control without mediator. Similar results were observed with regard to localized corrosion. Thus, the deepest pit observed on steel exposed to the media without addition of mediators had a depth of 10.4 μm. But when mediators were added, deepest pit was observed to have a depth of 22.2 μm when 10 ppm riboflavin was added and it was 20.2 μm when 10 ppm of FAD was added in the inoculated medium. It was almost the doubling of the pit depth due to addition of electron mediators. From industrial application standpoint of this observation, the implication is that a carbon steel pipe will experience pitting failure in half of the time when compared with the case when no electron mediators are present in the medium.

Consider a metal, e.g., iron immersed in an inoculated media. It will oxidize and release electrons to experience MIC. However, these electrons must reduce some oxidants inside/outside cell so that (i) the oxidation and hence corrosion of iron continues and (ii) the available energy as a result of oxidation/reduction or redox reaction helps in cell growth. The extracellular electron transfer (EET) in a fluid media, e.g., in case of biofilm formed over a surface, could be a metal surface, is difficult to materialize since it is not an electrical conducting media. It, therefore, needs to look for oxidants in the media which accept the electron as a consequence of a reduction reaction. The reduced oxidant which is no more an oxidant, transfers this electron through oxidation under suitable condition to another nearby oxidant. In this way, the electron moves through a chain of oxidants by experiencing a chain of reduction–oxidation reactions and finally reaching on cell wall. On approaching cell wall, it moves inside cell just like direct electron transfer. Inside the cell, the electron triggers reduction reaction, e.g., reduction of sulfate, nitrate, etc., with a high rate due to biocatalysis. These set of events lead to MIC of iron with a significant corrosion rate. This chain of oxidants is termed as shuttle for electrons which are 2–8 nm wide and therefore termed as conductive 'nanopili.' Each subunit of a pilus is a molecule of **protein**. These pili are attached to cell wall and give the cell a hairy appearance. A bacterial cell may have one pilus, several pili but most have typically between 100 and 1000. These pili conduct electrons from metal surface to the respiratory system (using the electron transport system) onto the final electron acceptor inside the cell. **This makes nanowires among the smallest known electrical wires and they were developed by nature long before human beings discovered electricity.**

The importance of nanopili-based EET can be appreciated if we follow the sequence of events that might be taking place when metal is exposed to an inoculated media and MIC starts and corrosion progresses. Let us consider a biofilm community at iron metal surface which presents an environment having high cell density which permits the buildup of the shuttle or a nanopilus. This closed system is expected to consist of many shuttles where reduction oxidation reactions could occur and an abundant external oxidizing iron metal. It can be predicted that mechanisms employed for energy generation (involving oxidation reduction reaction) at different layers of biofilm are expected to be different even for a single species biofilm. Let us consider metabolism in three bacterial cells in a media at three different places (Fig. 2.2): (1) Cell 1—at the metal surface, (2) Cell 2—in the middle of biofilm, and (3) Cell 3—on the top of the biofilm. First, let us consider situation when biofilm just starts building. In this colonization stage, the cells form a monolayer (involved cells are Cell 1 type) prior to establishing a three-dimensional structure (equivalent to multilayers one over the other). In this situation, the reduction reactions are primarily dominated by proteins which directly contacts the metal surface. In this condition, direct electron transfer (DET) is the dominant mechanism of energy generation through redox reactions. If there are any shuttle, responsible of extracellular electron transfer (EET) they are easily lost by diffusion into the immediate surrounding media. As the biofilm develops, the effective cells are Cell 2 type. One expects the local shuttle concentration to rise. This is also due to exopolysacharides (EPS) matrix, the constituents of biofilm, which are known to slow down diffusion of small molecules between the biofilm and directly contact the metal surface, for energy generation. Presuming anaerobic conditions existing at this level of biofilm (due to low oxygen solubility and diffusibility, and rapid oxygen utilization by aerobically respiring cells), the effective oxidants are SO_4^{2-}, NO_3^- in the presence of SRB, NRB, respectively, Fe^{3+} ions in the presence of iron-oxidizing bacteria, etc. Anthraquinone-2,6-disulfonate (AQDS), 2-amino-3-carboxyl-1,4-napthoquinone (ACNQ), etc., are redox shuttles which involve in EET. These events are responsible for energy generation for bacterial growth, alongside continuous oxidation of metal and its deterioration by corrosion (MIC). In case of cells like Cell 3, electron transfers are dominated by aerobic respiration. From the above discussion, it can be realized that most cells are C2 type because these are observed in most part of thickness of biofilm as compared to C1 and C3 cells.

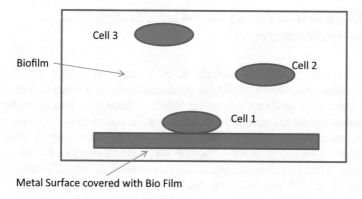

Fig. 2.2 Cells at three different places in a biofilm formed over a metal surface

Since for C2 cells neither the proximity of metal surface (as in case of C1 cells) nor the presence of oxygen (easily available oxidant) (as in case of C3 cells) is feasible, they have to make use of either the nanopili electron shuttles or mediated electron transfer for EET to materialize biocatalytic reduction, a requirement for energy generation for bacterial growth as well as continuance of metal corrosion.

2.4   Carbon Source Starvation

Bacteria, e.g., SRB, needing small amount of energy, for enzyme activity, do not need to corrode carbon steel for their survival if there is enough organic carbon atoms, required for their growth, accessible to them as electron donors. These carbon atoms are already present in culture medium, e.g., sodium lactate in modified Baar's medium for SRB also termed as ATCC medium. Under such circumstances, the bacteria do not require to corrode carbon steel for their survival. However, with continuous growth of live bacteria, the amount of these carbon atoms goes on decreasing leading to a situation termed as 'carbon starvation.' This phenomenon was demonstrated by experiments conducted on steel corrosion by *Desulfovibrio vulgaris* [9] biofilm which showed 10 μm maximum pit depth when carbon reduction was 90% while 99% carbon reduction yielded the highest specific weight loss of 0.0019 g/cm^2. Thus, one observes triggering of more aggressive corrosion under the condition of carbon starvation.

2.5 Biofilms in Inducing Corrosion Through Change in Electrochemistry

Biofilms formed by different bacteria, differ in their composition due to various metabolic reactions. Consequently, the products formed as a result of these electrochemical/chemical reactions may differ. Some of these may lead to enhancement in corrosive conditions and some in inhibition in the vicinity of metal surface. As such these effects, being bacteria specific, can be elaborated by considering different microbes independently. Bacteria which are mostly found to affect corrosion of metals and therefore are responsible for corrosion failures of industrial equipment formed of cast iron, mild and stainless steel, brass, etc., are sulfate-reducing bacteria, sulfur-oxidizing bacteria, iron-oxidizing/reducing bacteria, manganese-oxidizing bacteria, bacteria-secreting organic acids, and exopolymers or slime, etc. A brief discussion about them is given below.

- Sulfate-Reducing Bacteria (SRB)

These are perhaps the most studied bacteria from the point of view of microbial induced corrosion. This is, to a large extent, due to their wide spread presence in various type of environs, e.g., oil well, underground pipeline, seawater, river beds, various industries dealing with sulfur products, etc. Consequently, they are observed to have the largest impact on corrosion also because sulfate is widely distributed in anoxic environments. Also, corrosion effects of these bacteria on various material of construction is significant. As the name implies, these bacteria are sulfate-reducing type. That is they need presence of sulfate ions in the solution in order to survive. In addition to sulfate, SRB can also use other sulfur compounds with a valence above -2 as terminal electron acceptors. They include bisulfite ($HSO_3{}^-$), thiosulfate ($S_2O_3{}^{2-}$), and elemental sulfur. SRB's are anaerobic although some species of SRB's have been reported which are found to survive under low dissolved oxygen concentration. If we consider a solution without SRB, a set of reactions (1.1, 1.5a, b or 1.6) are expected to follow, which lead to corrosion of iron (steel). If we consider a solution, which is abiotic, corrosion is expected to be of significance if the solution is acidic and/or it is having dissolved oxygen. However, if we consider a solution which is near neutral (pH ~6–7) and is anaerobic, extent of corrosion on steel is not worth consideration. Since SRB's are anaerobic bacteria and they thrive under near neutral conditions, a very high degree of corrosion (~30–40 times or even more under certain conditions) may be observed in the presence of SRB's, in near neutral solutions containing $SO_4{}^{2-}$ ions. This observation hints at some special reaction, in the presence of bacteria, which results in enormous extent of microbial corrosion on metals. To understand this observation, several suggestions have been put forwarded, but final picture is yet to emerge out.

Most acceptable theory by far considers reactions (1.1)-iron oxidation, (2.1)-reduction of sulfate to sulfide and (2.5)-formation of iron sulfide:

$$Fe^{2+} + S^{2-} \rightarrow FeS \tag{2.5}$$

Since, the solution is anaerobic, one cannot consider oxygen reduction and it is near neutral solution so hydrogen evolution as reduction reaction is also ruled out. Thus, increased corrosion in the presence of bacteria was attributed to sulfate-reducing reaction which cathodically polarizes the anode and thus results in corrosion of iron. The sulfate reduction reaction was associated with the scavenging of hydrogen because of the ability of bacterial cultures to consume cathodic hydrogen and the stimulation (depolarization) of cathodic reaction resulting in iron corrosion. This theory was famously termed as 'cathodic depolarization theory' and was acceptable for quite some time. With passage of time, it was observed that no culture-based experiment was able to demonstrate that bacterial consumption of cathodic hydrogen accelerates iron corrosion to any significant extent. So, the previously observed acceleration of cathodic reactions in SRB cultures could now be explained by reaction between sulfide and iron rather than by microbial consumption of cathodic H_2. Now, it seems that much of the corrosiveness of SRB could be attributed entirely to their formation of H_2S, which is a powerful cathodic and anodic reactant. H_2S is known to rapidly react with metallic iron [net reaction, $H_2S + Fe^0 \rightarrow FeS + H_2$, $\Delta G° = -72.5$ kJ mol/Fe^0] thereby forming the characteristic corrosion product iron sulfide. Such biogenic iron sulfides is found to accelerate corrosion when deposited on the metal. Interestingly, sustained corrosion by iron sulfides required the presence of active populations of SRB. Until recently, SRB-induced corrosion was viewed as the result of biogenic H_2S and the catalytically active iron sulfides that are formed in the process of 'H_2S corrosion.'

Experimental evidence for a novel corrosion mechanism was furnished through isolation of SRB from enrichment cultures with metallic iron as the only electron donor [10]. Apparently, sulfate reduction by these peculiar strains was directly fueled by bacterial consumption of iron-derived electrons, without the involvement of cathodic hydrogen gas as an intermediate. In fact, while even the most efficient hydrogen-utilizing SRB did not accelerate iron corrosion compared to sterile tests when grown in organic matter-free (lithotrophic) cultures, these novel isolates accelerated iron oxidation up to 71-fold under the same conditions. The existence of such a direct mechanism of electron uptake had previously been considered by some investigators but without the availability of defined model organisms for experimental validation. Recently, the process was able to be studied in greater detail and the term 'electrical microbially influenced corrosion' (EMIC) was proposed. EMIC, which is fundamentally different from the corrosive effects of biogenic H_2S, can destroy metallic structures at rates of high technological relevance. Figure 2.3 shows the corrosion of an iron key explained in terms of electrical microbially influenced corrosion (EMIC) in the presence of Desulfovibrio ferrophilus (A to C) and corrosion under sterile (control) conditions (D to F). Both incubations were performed in artificial seawater medium at pH 7.3 and without addition of organic substrates (lithotrophic medium). Fig. (A) shows Bar, 1 cm. Iron key before incubation with D. ferrophilus strain IS5. (B) Iron key with biogenic corrosion crust after 9 months of incubation with pure culture of strain IS5.

(C) Residual iron after removal of the crust (B) with inactivated acid (10% hexamine in 2 M HCl) revealed 80.3% (2.7 g) iron weight loss due to corrosive activity

Fig. 2.3 Corrosion of iron key in presence of *Desulfovibrio ferrophilus* (**a–c**) and corrosion under sterile conditions (**d, f**)

of strain IS5. Hexamine-HCl did not dissolve Fe0. (D) Iron key before sterile incubation. (E) Iron key incubated in sterile artificial seawater medium. Corrosion is much less pronounced despite 27 months of incubation. (F) Residual iron after removal of corrosion products with inactivated acid (10% hexamine in 2 M HCl) revealed 2.9% (0.09 g) iron weight loss due to abiotic corrosion. While EMIC has so far been observed in only a limited number of highly corrosive SRB isolates, all SRB—by definition—can influence corrosion through excretion of the chemical H_2S ('chemical microbially influenced corrosion'; CMIC) if sulfate and suitable electron donors are present. In conclusion, SRB act as either direct or indirect catalysts of anaerobic iron corrosion (EMIC and CMIC, respectively) and there are species-specific differences in this respect.

- Iron-Oxidizing Bacteria (IOB)

Iron-oxidizing bacteria are another class of bacteria which have been found to affect MIC and further these bacteria are also found in various industrial environments, e.g., water tanks, pipelines, condensers, etc. These bacteria are aerobic microorganisms, belonging to a large and diverse group that get energy necessary for their metabolism from iron oxidation. Consequently, there is the formation of iron hydroxides that generally form insoluble precipitate on the surfaces, promoting regions with different oxygen levels. They are characteristically difficult to be isolated and cultured in the laboratory, being widely found in water from rivers, lakes, and oil production. They have mostly a locomotor sheath and their presence can be detected by a large accumulation of ferric precipitated as corrosion product. This accumulation or inorganic fouling leads to problems to industrial equipment such as blockages in oil pipelines. Among the most common iron bacteria are the species *Thiobacillus ferrooxidans* and the genera *Crenothrix*, *Gallionella*, *Leptothrix*, and *Spherotillus*. Various studies on MIC have established that extracellular polymeric substances (EPS) which are released by the bacteria, and therefore form a significant proportion of solid products in biofilm, play an important role in enhancing or inhibiting corrosion whether chemical or MIC. It also promotes the growth of sessile bacteria. In one of the studies, therefore, the effect of EPS on affecting corrosion on mild steel was studied by varying (i) EPS formed in 7, 14, and 28 days, (ii) amount of EPS in nutrient media, e.g., 24, 240, and 1200 mg/L, and (iii) effect of enzyme activity. It was observed that the corrosion was minimum for 7 day old EPS in a concentration of 240 mg/L. Further, heat deactivation of enzyme, component of EPS, gave better protection against corrosion. This behavior was related with the compactness of the EPS coating of the steel, however, more work related to structural investigation of EPS component formed in case of EPS of differing age and electrochemical reactions involved in them needs to be done to clarify the role of EPS on affecting MIC. Also, one needs to further investigate the role of denaturing (loss of bioactivity) of EPS enzyme on MIC since this gives us another green technology method of corrosion inhibition.

- IOB + SRB Inoculated Media

In presence of both IOB and SRB, the suggested MIC process involves several stages—(i) IOB's lead to forming of slime where there is low velocity or water is stagnant. (ii) The slime forming results in concentration oxygen cell with higher concentration of oxygen outside the slime and low concentration inside, (iii) dissolved Fe^{2+} formed as a consequence of iron oxidation inside the slime forming space moves outward toward tubercle since outer part is negative (cathode) as compared to the inner part (anode). While moving, Fe^{2+} ions are oxidized to Fe^{3+} by IOB's (*Gallionella*). Fe^{3+} form $Fe(OH)_3$. Wall of tubercle now consists of $Fe(OH)_3$ + slime + other bacterial species, if present. As written earlier, exterior of tubercle is cathodic while metal pit surface is highly anodic.(iv) As tubercle matures, slime/biomass begins to decompose, forming sulfates that attract SRB (sulfate-reducing bacteria), which then produce S_2^- and then H_2S in the interior. FeS is also possible to form. The related reactions are (2.1) and (2.5).

(v) Finally, if chlorides are also present with the Gallionella, Fe chlorides may form which are highly hydrolizing type hence on hydrolysis produce HCl, a highly acidic media. Thus, overall, in the presence of both IOB and SRB, the whole system encounters a situation having iron anode and outer part of tubercle as cathode with large potential difference due to acidic media and smaller anode and large cathode area. All these results into conducive environment for severe pitting attack of the metal pit. One observes deep pits on iron surface in such cases. For this purpose, corrosion attack was checked on X65 carbon steel (0.03 C, 0.17 Si, 1.51 Mn, 0.02 P, 0.17 Ni, 0.04 Cu, 0.16 Mo, 0.06 Nb, 0.02 Al, 0.01 Ti, and balance Fe as wt%) exposed to sterile and mixed SRB + IOB media for an exposure time of 5,13, and

21 days [11]. The cleaned corroded sample showed uniform corrosion and no pits in case of sterile media even for exposure up to 21 days, whereas in case of exposure in media inoculated with SRB and IOB, pits start appearing after an exposure of 13 days while very deep pits were observed in case of exposure to 21 days.

- Nitrate-Reducing Bacteria (NRB)

In the oil and gas industry, nitrate is sometimes injected to promote NRB growth to suppress SRB growth for the mitigation of reservoir souring through competition with sulfate reduction by SRB. However, iron oxidation coupled with microbially catalyzed nitrate reduction is thermodynamically favorable, even more so than iron oxidation coupled with sulfate reduction. A Bacillus licheniformis biofilm grown on carbon steel was found more corrosive than a sulfate-reducing *D. vulgaris* biofilm during a 7-day laboratory test. Nitrate-reducing *Pseudomonas aeruginosa* corrosion was reported on 304 stainless steel as well. Thus, nitrate injection should be metered carefully to prevent nitrate from entering pipelines. Unlike sulfate, unpolluted seawater does not contain nitrate. However, mixing of agricultural water introduces nitrate to water systems and soils. In recent years, researchers have realized that NRB MIC is as important as SRB MIC in soil MIC investigations.

It has been shown [7] that mediated electron transfer increases corrosion in the presence of *Psuedomonas aeruginosa*. In a case of EET in case of NRB *Pseudomonas Aeruginosa*, addition of 10 ppm riboflavin increased the weight loss from 2.06 to 2.34 mg/cm^2 and on adding same amount of FAD to 2.61 mg/cm^2. The addition of electron transfer mediators do not increase planktonic cells but they increased the concentration of sessile cells, in the media, slightly since they are benefitted from the energy increase due to accelerated MIC. One also observes increase in extent of pitting due to this change. Thus, 1018 carbon steel experiences pitting (maximum pit depth 17.5 μm) in media with P.aeruginosa, more sever pitting is observed in case of exposures to media having P.aeruginosa with 10 ppm FAD (max pit depth 25 μm) and riboflavin (max pit depth 24.4 μm), respectively. The average pit depths in the three cases were 16.2 ± 1.3, 23.2 ± 1.8, and 21.9 ± 2.5 μm, respectively.

2.6 Fungi

In warm and humid climates such as Southeast Asia, fungi become an important factor in MIC investigations. Fungi are among the most common microorganisms found in the air, soil, foodstuffs, paint, textiles, bird feathers, and on live and dead plants. Most fungi are capable of producing organic acids and are implicated in the corrosion of steel and aluminum, especially in failure of aircraft fuel tank, and also magnesium alloy and zinc. Thus, (i) **Cladosporium resinae** are involved in the corrosion of aluminum alloys. Cladosporium is the most common member of the so-called black molds. This fungus breaks down aromatic ring compounds to simpler hydrocarbons and utilizes them for its metabolism. It produces organic acids by metabolizing fuel components in aircraft fuel tanks, leading to fuel breakdown and corrosion of fuel tank. (ii) The fungus **Hormoconis resinae** utilizes the hydrocarbons of jet fuel to produce organic acids. Surfaces in contact with the aqueous phase of fuel–water mixtures and sediments are common sites of attack. The large quantities of organic acid by-products excreted by this fungus selectively dissolve or chelate the copper, zinc, and iron at the grain boundaries of aircraft aluminum alloys, forming pits which persist under the anaerobic conditions established under the fungal mat. (iii) **Aspergillus niger** and **Penicillium spp.**, both fungal species are known to produce citric acid which may be involved in the attack on grease-coated wire rope wound on wooden spools stored in a humid environment.

Secretion of organic acid occurs during fermentation of organic substrates. The impact of acidic metabolites is intensified when they are trapped under the biofilm due to hydrolyzing nature of iron compounds which results in increased acidity of the media. Organic acids produced, e.g., acetic acid, oxalic acid, etc., increase the corrosion of aluminum depending upon pH value. The inhibitory effect of acetate and oxalate anions in neutral media can be explained due to the formation of a precipitation type compound. However, in the acid solutions, pitting occurs at more negative values of pitting potential than in neutral chloride solutions making the aluminum alloy susceptible to pitting. The acidity can prevent repassivation process facilitating crevice corrosion as well.

Formation of oxygen concentration cell is also responsible for enhanced corrosion in the presence of fungi. Due to restriction of oxygen diffusion underneath the mature biofilm, the concentration of oxygen under the biofilm depletes significantly. This leads to formation of concentration cell with practically no oxygen beneath the biofilm and normal oxygen concentration in the uncovered zones of metal surface (Fig. 2.4).

Area of metal surface under the biofilm acts as anode while the uncovered area acts as cathode leading to severe corrosion attack on metal under the biofilm, which sometimes also appear as pit.

Other reasons for fungal attack could be (i) production of anaerobic sites, due to fungal growth, for SRB and metabolic by-products that are useful for growth of various bacteria; (ii) production of a surfactant by C. resinae that degrades aircraft fuel by allowing water to partially mix with it, creating an emulsion which affects the combustive qualities of the fuel, ultimately resulting in fuel failure and machine damage; (iii) isolation of iron-reducing fungi from tubercles in a water distribution system, suggesting another mechanism, whereby corrosion may be accelerated by

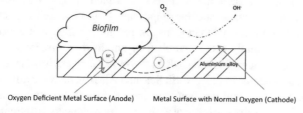

Fig. 2.4 Oxygen concentration cell

this group of microorganisms; (iv) the microorganisms introduced into fuel tanks from airborne contaminants and through water enters the tank. Consequently, MIC is observed on aluminum alloys AA2024 and 7075, used in aircraft or in underground fuel storage tank.

A recent work [12] reports fungal-induced corrosion on steel, aluminum etc., exposed to hydrocarbon fuels during transportation or storage and so is of significance for aircrafts. Tests done on corrosion of aluminum due to Acrimonium kiliense, a fungus, shows increase in corrosion rate from 0.53 mpy for an exposure of 10 days to 2.00 mpy for an exposure of 60 days. For exposure in control media, corrosion rate is ~0.53 mpy and remains more or less unchanged upto 60 days. The increase in corrosion rate with time has been suggested to be due to accumulation of metabolites under fungal colonies attached to the metal surface and to the oxygen concentration cells.

2.7 Algae

Algae and eukaryotic organisms are ubiquitous in nature. They are present in various sizes, shapes, and range from unicellular to multicellular forms. Algae are autotrophic and derive their energy from carbon dioxide, water, and sunlight. Algal growth results in drastic changes in dissolved oxygen and pH in a water body. The general classification of the algae is based partly upon the nature of the chlorophylls and accessory pigments, Chlorophyta (green algae), Rhodophyta (red algae), and Phaeophyta (brown and other pigmented algae). In industrial units, algae generally flourish on wetted, well-lit surfaces such as cooling towers, storage tanks, and distribution systems. Due to their capability to generate oxygen, organic acids, and nutrients for other organisms, algae play an indirect role in microbial proliferation and deterioration of materials. The corroding action of the algae is very slow and results in very thin layer of corrosion products. Algae also promote corrosion of materials by carbonic acid formed from carbon dioxide liberated during their respiration.

2.8 Archaea

In extreme environments, such as reservoirs in which temperature can easily reach 70 °C or higher, archaea become important. Some archaea are sulfate or nitrite reducers and some are methanogens. In recent decades, it has been realized that archaea are actually more abundant in nature than previously thought. Thermophilic sulfate-reducing archaea (SRA) thrive in many environments such as marine hydrothermal systems in Italy, North Sea, and Alaskan oil reservoirs. Just like SRB, SRA also cause MIC and reservoir souring because they both use sulfate in their metabolism [1]. Also, the existence of archaea sulfate reducers at extremely high temperatures could explain hydrogen sulfate formation in hot sulfate-containing environments and so may also be responsible for MIC of pipelines and other machinery parts in deep oil wells.

References

1. Jia R, Unsal T, Xu D, Lekbach Y, Gu T (2019) Microbiologically influenced corrosion and current mitigation strategies: a state of the art review. Int. Biodeterior Biodegradation 137:42–58
2. Huang L, Regan JM, Quan X (2011) Electron Transfer Mechanisms, new applications and performance of biocathode of microbial fuel cells. Biores Technol 102:316–323
3. Hernandez ME, Newman DK (2001) Extracellular electron Transfer. Cell Mol Life Sci 58:1562–1571
4. Kato S (2016) Microbial extracellular electron transfer and its relevance to iron corrosion. Microb Biotechnol 9:141–148
5. Li Y, Xu D, Chen C, Li X, Jia, R, Zhang D, Sand W, Wang F, Gu T (2018) Anaerobic microbiologically influenced corrosion mechanisms interpreted using bioenergetics and bioelectrochemistry: a review. J Mater Sci Tech 34:1713–1718
6. Zhang P, Xu D, Li Y, Yang K, Gu T (2015) Electron mediators accelerate the MIC of 304 stainless steel by the *Desulfovubrio vulgaris* biofilm. Bioelectrochemistry 101:14–21
7. Jia R, Yang D, Xu D, Gu T (2017) Electron transfer mediators accelerated the microbiologically influence corrosion against carbon steel by nitrate reducing *Pseudomonas aeruginosa* biofilm. Bioelectrochemistry 118:38–46
8. Li H, Xu D, Li Y, Feng H, Liu Z, Li X, Gu T, Yang K (2015) Extra cellular electron Transfer is a bootleneck in the microbiologically influenced corrosion of C1018 carbon steel by the biofilm of sulfate reducing bacterium *Desulfovibrio vulgaris*. Published on line 2015 Aug 26. doi: https://doi.org/10.1371/journal.pone.0136183
9. Xu D, Gu T (2014) Carbon source starvation triggered more aggressive corrosion against carbon steel by the *Desulphovibrio vulgaris* biofilm. Int Biodeterior Biodegrad 91:74–81
10. Enning D, Garrelfs J (2014) Corrosion of iron by by sulfate reducing bacteria: new views of an old problem. Appl Environ Microbiol 80:1226–1236
11. Lv M, Du M, Li X, Yue Y, Chen X (2019) Mechanism of microbiologically influenced corrosion of X65 steel in seawater containing sulfate-reducing bacteria and iron oxidizing bacteria. J Mater Res Technol 8:4066–4078
12. Imo EO, Chidibere AM (2019) Fungus Influenced corrosion of alumium in the presence of *Acrimonium kiliense*. Int J Appl Microbiol Biotechnol Res 7:1–6

Chapter 3
Microbial Induced Corrosion and Industrial Economy

Abstract This chapter deals with deleterious effect on industrial machinery and other material structure due to Microbial Induced Corrosion. Corrosion, due to only chemicals, weakens the material strength prematurely and thus results in unscheduled shutdown of plants, catastrophic failure pf machinery, leakage of chemicals etc. This needs maintenance of machinery or its replacement. Other factors are also involved which together results in extra expenditure on industry to keep the system operational. This is Corrosion cost and it affects the economy of industrial production. Similar effect is also observed with MIC. The chapter discusses various aspects related to failure of machinery and the extent to which MIC is observed in various industries or structures etc. Although, it has not been possible to systematically evaluate damage due to MIC due to reasons e.g. (i) difficult to identify whether failure is due to corrosion or MIC or (ii) combining the effect of biocorrosion with biofouling etc. Still effort has been made to put an estimated cost due to MIC on the basis of data present in literature on MIC. An account of various industries and their various sections experiencing MIC has been given which helps in appreciating the deleterious effect of MIC on machinery, vehicles and structures etc.

Keywords Direct cost · Indirect cost due to MIC · Corrosion cost · Industry sectors · Cost estimate

Corrosion is a phenomenon which weakens material of construction be it a bridge, building, railway track, aircraft, armaments, etc., in the civil and defense sector, industrial machinery, e.g., pipelines, storage tanks, reaction vessels, high-pressure vessels, filters, boilers, valves, pumps, etc. Through this effect, it is responsible for malfunctioning of these equipment leading to loss of chemicals, materials, heat, power, etc., and in extreme conditions, for premature failure and breakdown of structures and machinery which in turn shortens their useful life and many times catastrophic failures resulting in loss of life of the personnel working in the plant. Leakage of chemicals from plant, due to corrosion failures, may result in pilferation of toxic and poisonous chemicals in environment leading to serious air and water pollution. This in turn may affect human life. An eye opener case is Bhopal Gas Tragedy, Bhopal (India) which happened on the night of December 2, 1984. Failure of local plant machinery led to leakage of toxic methyl isocynate gas from a chemical plant, inhalation of which resulted in death of over 4000 people. It can be seen from following examples [1] that corrosion causes too many accidents each year. These examples are: (i) On November 22, 2013, the Donghuang II oil **pipeline** suddenly exploded in Qingdao in eastern China. The blast killed 62 people and injured 136. (ii) In 2009, a 50-foot-tall, **high-pressure vessel** at a Crystal manufacturing facility in Belvidere, Illinois, exploded, injuring bystanders and killing a trucker at a nearby gas station. (iii) On May 20, 2000, **a pedestrian bridge**, an 80-foot section of the concrete and steel walkway snapped in half. Pedestrians fell 17 feet to the highway below. The bridge failure injured 107 people, 13 critically. (iv) on August 19, 2000, a 30-inch **natural gas pipeline** exploded, leaving an 86-foot-long, 46-foot-wide and 20-foot-deep crater. Twelve people died in the 1,200-degree fireball. The cause of all of these accidents was pipe or structural failure due to corrosion. Additionally, agriculture products and aquatic life, e.g., fish, prawns, tortoise, crocodile, etc., are also affected.

To avoid these unwanted happenings, one has to (i) carefully select an appropriate material of construction from a whole gamut of materials available commercially in the market from the view point of not just corrosion and wear resistance in the presence of chemical environment, high temperature, and pressure but also strength, hardness, toughness, high temperature stability, ductile character at low to sub-zero temperatures, machinability, and fabrication, etc., and above all cost of material; (ii) maintain an inventory of spare materials in the form of pipes, sheets, plates, rods, etc., so that the failed part of the machinery may be replaced in the shortest possible time, otherwise, downtime of plant will increase which enhances the cost of production due to reduction in-plant output; (iii) have a maintenance department which can take up the jobs as mentioned in the preceding point; (iv) install monitoring instruments for measuring process parameters, extent of corrosion, etc., so on and so forth. All this result into enhanced cost of installation and running of a production plant. They all fall under the category of 'corrosion cost' if the above precautions are practiced in a plant where operating conditions and environment may be very corrosive in nature, e.g., a chemical industry, process industry, oil and natural gas exploration, transport industry, etc. In turn, it affects the economy of a production from these plants or operating these amenities. To have an idea of corrosion cost or effect of corrosion on industrial economy, corrosion audits have been conducted by various agencies, from time to time, on different industrial sectors. These audits help to get an idea about the ongoing practices being carried out in the industry to minimize the corrosion cost and how far they are able to serve the purpose. If there are any lacunae or shortcomings in the practices, in vogue, one may feel the need to suggest carrying research in the area of material development, or on aspects related to corrosion science and technology or in altering the process conditions/practices so as to improve the economy of production which otherwise was getting affected due to corrosion.

Keeping the above in notice, one may look at the results on the basis of corrosion audit carried by different agencies in recent past. This gives a comparative idea about corrosion cost incurred by different industries. While calculating corrosion cost, one considers direct and indirect cost due to corrosion as given below.

Table 3.1 Cost of corrosion in billion US$

Country/regions	Cost	% GDP
USA	451.3	2.7
India	70.3	4.2
European region	701.5	3.8
China	394.9	4.2
Russia	84.5	4.0
Japan	51.6	1.0
Global	2505.4	3.4

Direct Cost:

- Cost of design

   Materials selection, coating, sealants, inhibitor, and cathodic/anodic protection,
   Manufacturing and construction, labor cost, transportation, and installation

- Cost of management

   Inspection
   Rehabilitation
   Repair
   Loss of productive maintenance

Indirect cost:

- Loss of productivity because of outages, delays, and failures,
- Litigation,
- Taxes of the overhead corrosion cost.

Table 3.1 gives an estimate of corrosion cost (in billion US$) [2] incurred by some of the advanced industrial countries/regions:

From above table, one can see that annual corrosion cost ranged from appx. 1–5% of GDP of each nation. A seminal study of 2004 estimated the annual direct cost of corrosion at $276 billion—about 3.1% of the US gross domestic product at the time. It reveals that, although corrosion management has improved over the past several decades, but there is still urgent need to find more and better ways to encourage, support, and implement optimal corrosion control practices [3]. These implementations lead to reduced cost and hence better economy in normal life too. One can understand this by considering one of the biggest success stories in the world when it comes to corrosion control—the automotive industry [1]. This industry is one which recognized the cost of corrosion quite early and worked actively to figure out solutions to control it. Before 1975, car manufacturers used minimal corrosion control. The life cycle of the cars these companies created often depended on the corrosion of the body. And, corrosion was costing the auto industry billions of dollars annually.

However, after 1975, the world's top car manufacturers began applying corrosion control measures, by using advanced painting/coating technology and corrosion-resistant materials. By 1999, it was estimated that the auto industry was saving $ 6.9 billion annually. Further, average age of vehicles increased by nearly 50% from 1975 to 1999. A win–win situation for autocompanies as well as for the consumers.

It should also be emphasized that in a modern business environment, successful enterprises cannot tolerate major corrosion failures, especially those involving **personal injuries, fatalities, unscheduled shutdowns, and environmental contamination.** For this reason, considerable efforts are generally expended in corrosion control at the design stage and in operational phase. Hence, a large amount of money can be saved by shifting paradigm of 'corrosion as maintenance issue' to a new paradigm '**corrosion control as integrated company plan**.' Therefore, it is critical that organizations come up with comprehensive corrosion management systems. The cost of industrial rust is enormous, but with corrosion management system in place, it is estimated that between 15–35%, which is, respectively, 375–877 billion US$, of the financial cost of corrosion can be mitigated worldwide. This is no small achievement when we are talking about numbers that are equal to more than 3% of the worldwide GDP.

In order to achieve goals mentioned above, although technological advancements have provided many new ways to prevent corrosion, better corrosion management can be achieved using preventive strategies in non-technical and technical areas. These strategies are:

- Change the misconception that nothing can be done about corrosion.
- Increase awareness of significant corrosion costs and potential cost savings.
- Change policies, regulations, standards, and management practices to increase corrosion cost savings through sound corrosion management.
- Improve education and training of staff to enable them recognize importance of corrosion control.
- Implement advanced design practices for better corrosion management.
- Develop life-prediction and performance-assessment methods.
- Improve corrosion technology through research & development and check their utility by scrutinizing their Pilot scale and in-plant test results.

As indicated earlier, corrosion can cause dangerous and expensive damage to everything from automobiles, home appliances, and drinking water systems to pipelines, bridges, and public buildings. Any study on estimating corrosion cost has to be conducted considering various industry categories since the working environment and hence the extent of corrosion that materials are likely to experience will differ for different industry. In one such case of corrosion audit, first industry categorization was done and then corrosion cost was calculated industry sector wise. Thus, all the industries were divided into five categories (i) **infrastructure**, e.g., buildings, bridges, pipelines for gas and liquid including those providing potable water, for carriage of crude, etc., roads, plants, pipelines, tanks, water ways, ports, airports, rail road, etc. Corrosion is typically found on piers and docks, bulkheads and retaining

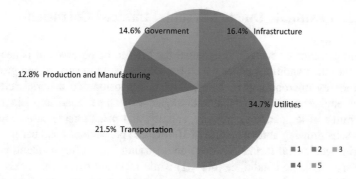

Fig. 3.1 Corrosion cost (%) in various industry sectors

walls, mooring structures, and navigational aids and other key elements of the infrastructure. Today, as much of the aging infrastructure reaches the end of its designed lifetime, the emphasis is on maintaining and extending the life of these valuable assets. (ii) **Utilities**, e.g., supply lines for gas, drinking water and sewage system, electricity, and telecommunication; (iii)**transportation**, e.g., motor vehicles, ships, aircraft, rail roads, hazardous material transport, etc.; (iv) **production and manufacturing**: This category includes industries that produce and manufacture products of crucial importance to the economy of a nation and for improvement of standard of living of its residents. These include oil production, mining, petroleum refining, chemical and pharmaceutical production, and agricultural and food production; (v) **government**: Defense, e.g., corrosion of military equipment and facilities particularly more important due to (a) slowing down of acquisition of new equipment, (b) possibility of these equipment operating under harsh environmental conditions, e.g., too humid and salty, hot or cold and rocky surface, and nuclear waste storage for effective shielding from radiation emitting out of waste. Figure 3.1 shows % wise distribution of the corrosion cost incurred by various industry sectors. The distribution shows that various sectors show effects of corrosion to considerable extent with maximum for utilities sector probably because nearly all population utilize one of the utilities named here, although corrosion may not be severe in many of these utilities. On the other hand, many of the production units/process may be having very corrosive environment utilizing state of the art corrosion mitigation measures yet they show least % of corrosion cost incurred by production and manufacturing sector probably because total cost spent in this sector may not be as much as in utility sector.

All the above-discussed industry sectors and their subsectors experience considerable microbial induced corrosion particularly where (i) the existing conditions are not aggressive enough to cause significant normal corrosion to metallic structures, (ii) the environment is conducive for the growth of bacterial activity, and (iii) the possibility of biofilm forming on a metallic surface which could be a part of the infrastructures, etc.

3.1 Cost Estimate Due to Microbial Induced Corrosion

Since the presence of bacteria is essential for MIC to be observed, it is necessary to look into the conditions under which bacteria is likely to survive and grow [4]. Work done by microbiologists, bacteriologists, corrosion and material scientists, chemical engineers, etc., show that microorganisms can be found surprisingly over a wide range of temperature, pressure, salinity, and pH. Extreme cases could be, e.g., (i) SRB growing at temperature of 104 °C and pressure of 1000 bar in case of oil bearing geological formations deep underground, (ii) sulfur-oxidizing bacteria create very oxidizing conditions (pH < 1) while there are some which exist at the other end of pH scale, etc. Further, irrespective of the environmental conditions, microorganisms need water, source of energy to drive their metabolism, and nutrients e.g., C, N, and P, to provide essential food for cell renewal and growth. Energy usually is derived from sunlight, from photosynthesis or from electrochemical (redox) reactions. The latter form of energy generation is quite important for MIC and in most cases of MIC's, e.g., closed systems and buried facilities like underground pipelines, tanks, etc. Consequently, various materials of construction of steel, stainless steel, copper, and Al and their alloys, etc., are observed to be affected by MIC as a part of water handling operations and manufacturing processes in oil and gas production, pipelining, refining, petrochemical plants, power production, wastewater treatment, drinking water supply, pulp and paper making, etc. An idea of dominance of MIC over chemical corrosion can be had from an example of corrosion of pipes for transporting potable water system in one production unit. The metallic material of construction for these pipes is cast iron, ductile iron, and steel. Of these cast iron pipe was found to corrode maximum due to bacterial presence. To avoid occurrence of MIC, they used biocides which proves to be cost effective even though biocides are oxidizing in nature and cause enormous corrosion to cast iron but still it is lesser than MIC. So the use of biocides reduces overall corrosion of pipes and hence reduced corrosion cost.

Although importance of estimating cost attributable to MIC is emphasized in no uncertain terms by considering above industrial example. Never the less, due to difficulty in observation of MIC previously due to lack of appropriate experimental facilities and lack of understanding of the phenomena, it was difficult to estimate the extent of MIC particularly where it was thought to exist alongside chemical corrosion. Also, many people wrongly associated MIC with chemical corrosion. Moreover, often, financial losses due to damage of equipment by biocorrosion are combined with those resulting from biofouling. While the two phenomena may be associated with each other, they do not cause the same type of damage. Hence, the costs associated with MIC usually included the costs of prevention of both MIC and biofouling; since these are based on a limited understanding of the phenomena, they could be under or overestimated. Consequently, computed basis of estimates of the costs attributable to MIC of iron, etc., are lacking and numbers associated with this parameter runs the risk of showing wide variation and definite numbers cannot be given with certainty. In spite of this, microbially influenced corrosion (MIC) probably accounts for a significant fraction of the total corrosion costs, and, due to its effects on important infrastructure in the energy industry (such as oil and gas pipelines), costs in the range of billions of dollars appear realistic. The contribution of MIC to this cost is ~50%. Industries most affected by MIC are power generation, oil exploration, transportation and storage and water distribution and in general industries with marine environment.

At this stage, one can talk about a few estimates.

1. As per one estimate, about 20% of all damages caused by corrosion are influenced by microorganisms; corrosion is the main cause of problems in the pipes of oil industry, affecting the costs of production and storage.

2. Another estimate suggests that 40% of pipe corrosion for oil industry is attributed to microbiological corrosion and it causes havoc from the order of US$100 million in production, transportation, and storage of oil every year in many countries.

3. According to another estimate, ~30–78% of corrosion cost is due to MIC. Here, the overall corrosion cost has been estimated to be US$ 110.7billion/year. The costs associated with the repairs is ~US$ 70 billion/year. Thus, on average, one can easily associate tens of billion US$ expenditure per year to take care of losses due to MIC alone.

4. In a study on pipeline integrity, in an oil and gas transmission system, carbon steel was used for pipeline fabrication. In this case, external corrosion accounted for nearly 38% of the structural integrity along ~60,000 km of pipeline. In this, anaerobic bacterial corrosion was estimated to account for 29% of all examined corrosion sites.

5. Few more examples of MIC dominated corrosion cases are given below:

    (i)   One power provider company has detected MIC of carbon steel in cooling water systems in virtually all their power plants. The costs associated with repairs and down time were estimated to be millions of dollars annually.

    (ii)   Under-deposit pitting of heat exchanger tubing in nuclear power generating plants has been estimated to cost $ 300,000 per unit per day in replacement energy costs. Corrosion problems have cost the nuclear utility billions of dollars in replacement costs alone.

    (iii) Losses in the oil and gas industry are also substantial. It was estimated that 34% of the corrosion damage experienced by one oil company was related to microorganisms.

    (iv)   In one of the major economies, 50% of corrosion failures in pipelines have been suggested to be due to MIC, while another work proposed that approximately 20% of all corrosion damage to metallic materials is microbially influenced. Replacement costs for biocorroded gas mains were recently reported to be £250 million per annum.

6. Microbiological corrosion and biofouling of materials are two main reasons of marine corrosion to cause the damage and failure of the equipment and structure existing in marine environment. The MIC is caused by SRB, IOB, and

their biofilms. In case of marine/maritime corrosion, bacteria form multispecies biofilms with other microorganisms. Bacteria and microalgae are responsible for microfouling and allow the adhesion of larger organisms such as algae, mussels, and barnacles which cause macrofouling. They adhere to almost all surfaces immersed in seawater, such as pipes and ship hulls. The costs associated with biofilms attached to ship hulls are mainly due to increased fuel consumption resulting from higher frictional drag, although the costs for hull cleaning, coating, and painting may also be considerable. According to one estimate, annual corrosion-related cost for shipping industry, considering new construction, maintenance, repair, and corrosion-related downtime, comes out to be US$ 2.7 billion. Hull fouling for the medium size naval surface was estimated to cost ~US$ 56 million/year. Fouling of ballast water with non-indigenous species resulted in additional expenditure ~1.6–4% of annual operational cost for a ship.

3.2 Industries Affected by MIC

After discussing the severity of MIC and the billions of Dollars, the industry need to spend to overcome ill effects of MIC, one may be interested in knowing various sectors/industries where MIC is important to deal with to reduce the maintenance cost. These sectors/industries are shown in below given Fig. 3.2 and Table 3.2 lists industrial area susceptible for MIC and metals/alloys used in fabrication of various systems/machinery parts.

MIC has been noted to occur for a variety of metals and alloys such as iron and steel, copper, and titanium, as well as on non-metallic materials such as concrete. MIC and associated biofouling are estimated to account for 20% of direct fuel pipeline

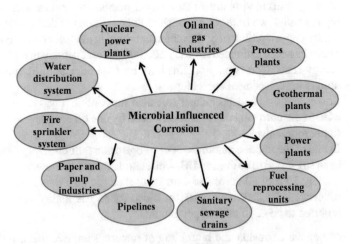

Fig. 3.2 Microbial induced corrosion in various industries

Table 3.2 Some examples of systems affected by microbial influenced corrosion

System/application	Problem components/area	Microorganisms
Maritime transport	Ship hulls, pipes immersed in seawater	Mussels and barnacles responsible for macrofouling, desulfovibrio and iron-reducing microorganisms. Aerobic microorganisms such as *Pseudomonas* sp.
Cooling system	Heat exchanger, cooling towers, storage tank	Aerobic (iron/manganese-oxidizing bacteria) and anaerobic bacteria (sulfate-reducing bacteria and sulfur-oxidizing bacteria)
Pipelines	Stagnant part of interior and external part of buried pipelines specially in wet environments	Aerobic (metal-oxidizing bacteria) and anaerobic (sulfate-reducing bacteria), slime forming bacteria, algae
Nuclear power generation plants	Condensers and heat exchangers, water pipes, and tubes	Aerobic (metal-oxidizing bacteria) and anaerobic (sulfate-reducing bacteria)
Fire sprinkler system	Stagnant areas	Anaerobic (sulfate-reducing bacteria) and aerobic (metal-oxidizing bacteria)
Vehicle fuel tanks	Stagnant area	Fungi
Oil and gas industries	Pipeline network and associated infrastructures	Sulfate-reducing bacteria

integrity and reliability costs exceeding $2 billion USD per year [5]. A brief description of problems faced in various sections of different industries has been given below.

- Heat Exchanger

In heat exchangers, various undesirable effects such as clogging of pipes and efficiency loss of heat exchangers, etc., have been observed due to microbes. The heat exchangers provide suitable conditions for the growth of microorganisms, e.g., SRB, IRB, *Pseudomonas* sp., and IOB. In one case of biofouling and corrosion in nuclear test reactor, it was observed that above 90% tube bundles of thermofluid heat exchanger (TFHX) were clogged due to biofouling and MIC. Microbiological assay showed the presence of SRB and IOB. Another study reported that the microbial reduction of nitrate to ammonia was responsible for the stress corrosion cracking of brass tubes in nuclear power plant.

- Water Distribution System

In case of water distribution system, it was concluded that biofilm produced by microbes was responsible for the deterioration of copper pipes. Another study reported the corrosion of cast iron due to SRB and IOB in water distribution system, and SRB demonstrated higher corrosion rate than IOB. In a case of MIC in cooling towers, water samples collected from them were analyzed to have corrosive bacterial community consisting of *Massilia timonae as* bacterial species showing maximum corrosive activity. In another study on the effect of biofilm on corrosion of iron pipes of drinking water distribution system, it was observed that biofilm accelerate deterioration of iron pipes initially but inhibit corrosion in later stage.

- Oil and Gas Industry

MIC is a serious issue in oil and gas industries [6]. It has been identified in several different locations in oil and gas assets such as oil and gathering pipelines, multiphase gathering stations, produced water handling process, seawater injection plant, gas pipelines, gas storage field line and crude storage tanks. MIC and internal corrosion in pipelines are closely related with the properties and composition of fluid transported by these pipelines and also operation conditions such as flow rate, temperature and pressure. Different type of bacteria including SRB was isolated from oil and gas industries which were responsible for the deterioration of metals. *Bacillus cereus* ACE4 strain, a manganese-oxidizing bacteria, was isolated from the oil and gas producing system and responsible for the corrosion process.

- Paper and Pulp Industry

MIC extends its dangerous signs to corrosion in paper and pulp industry too. In this industry, the blackening of paper and pulp was observed due to microbial activity of microbes specially SRB. SRB also enhance the disposal problem, which leads to the pollution, e.g., paper and pulp plant wastes were disposed to the river and leads to sulfide pollution. Slime problem are prominent in paper and pulp industries. In paper industry, corroded areas were found covered by slime and under the slime layer anaerobic bacteria were identified. Paper machine section is specially affected by MIC due to presence of cellulose, sulfate ions, neutral pH, and moderate temperature which provide ideal environment for growth of SRB's particularly under paper mat forming on machine wire.

References

1. White T, Cost of corrosion
2. Koch G, Varney J, Thompson N, Moghissi O, Gould M Payer G (2016) International measures of prevention, application and economics of corrosion technologies study. NACE International IMPACT, 1–72 (March 1, 2016)
3. Koch GH, Brongers MPH, Thompson NG, Viramani YP, Payer JH (2002) Corrosion costs and preventive strategies in the United States. NACE Publication No. FHWA-RD-01-156
4. Jack TR (2002) Biological corrosion failures, ASM Handbook, vol 11: failure analysis and prevention. ASM International 2002, pp 881–891

5. Beech IB, Gaylord CC (1999) Recent advances in the study of Biocorrosion—an overview. Rev De Microbiol 30:177–190
6. Geig LM, Jack TR, Foght GH (2011) Biological souring and mitigation in oil reservoirs. Appl Microbiol Biotechnol 92:263–282

Chapter 4
Experimental Details and Strategies to Control Corrosion

Abstract This chapter is divided into three parts. First two parts discuss about experimental details related to experiments which may be required to do in investigation on MIC. First of these is microbiology work, e.g., isolation of bacteria from natural resources, studying the growth cycle of bacteria, and biofilm characterization, and second of these is related to investigating corrosion due to bacteria. In this later case, details have been discussed about surface preparation of original and corroded metal samples, preparation of test solution, corrosion estimation test, e.g., immersion test, electrochemical test, analytical tests, e.g., XRD, FTIR, SEM, and ESCA. The third part deals with the strategies to control MIC. This part discusses about corrosion-causing bacteria, e.g., sulfate-reducing bacteria (SRB), iron-oxidizing bacteria (IOB), metal-reducing bacteria, and acid-producing bacteria, and the details involved which are responsible for MIC. Afterward, in this chapter, details about the bacteria which mitigate MIC or corrosion have been dealt. This part deals with several different ways in which microbial induced corrosion inhibition can be practiced using bacteria. Thus, microbial inhibition has been discussed (i) by removal of oxygen, (ii) by inhibiting growth of corrosion-causing bacteria through production of antimicrobials, and (iii) by using microbes which produce protective layers.

Keywords Growth cycle · Biofilm · Metal surface · Corrosion tests · Analytical tests · Inhibition

It has been discussed in the preceding chapters, in brief, about the (i) corrosion of engineering materials in the presence of bacteria, (ii) conditions which are conducive for the presence and growth of bacteria, and (iii) mechanism of MIC specific to the bacteria. We now take a look at these aspects in little bit more in detail including the information about the media which are conducive for the growth of bacteria, their composition, preparation, and precautions while preparing these solutions. Since now, one is familiar with the details with regard to corrosion fundamentals, the occurrence of MIC, etc., it will be logical to discuss about the ways as to how to control/mitigate these cases of MIC.

Accordingly, this chapter is divided into three parts: Sect. 4.1—Experimental details related to isolation, culture, and storage of bacteria either from standard/natural source, growth, characterization, and bacteria and biofilm, etc.

© The Author(s), under exclusive license to Springer Nature Singapore Pte Ltd. 2020   57
A. K. Singh, *Microbially Induced Corrosion and its Mitigation*,
SpringerBriefs in Materials, https://doi.org/10.1007/978-981-15-8019-2_4

Sect. 4.2—Experimental set-up and other details in connection with estimation of extent of corrosion and its type, analysis of corroded metal surface, corrosion products, etc., and Sect. 4.3—Brief about various corrosion-causing bacteria and strategies for controlling MIC.

4.1 Experimental Details—Bacteria/Biofilm

Any type of corrosion is recognized as microbial induced corrosion if it is observed on any metal/material surface due to the presence of bacteria in the conducting media such as any aqueous solution. That means if there has/have been no bacteria (i.e., aquatic media) in this media, the material may not have experienced any significant degree of corrosion because of mere presence of some chemicals in that media. This point probably needs some amount of explanation. It may be interesting to note that even if one takes distilled water, which is supposed to have minimal amount of chemicals, a metal like iron will experience some degree of corrosion because of the presence of dissolved oxygen which acts as oxidizing agent by reducing itself. But although corrosion resistance of iron is known to be very poor, this level of corrosion (as quantified in terms of 'corrosion rate') will not be high enough to reduce the lifetime of the iron to significant level so as to make it uneconomical to use (Chap. 3). It can be understood in terms of decrease in load bearing capacity of a material as its thickness decreases. Corrosion of a material reduces its thickness after some passage of time (consider metal in the form of a flat piece having certain thickness) which can be estimated by measuring 'corrosion rate' per year basis (Chap. 1). From this data, one can estimate reduction in load bearing capacity (from reduced thickness) after passage of certain period of time say 5 years, 10 years, and so on. Once thickness of material plate reduces below 'minimum required thickness,' the plate becomes unusable as its continuous use will result in its failure and so the plate needs to be replaced. This sudden failure is leading to unscheduled shut down which results in stoppage of production. All this adds up to corrosion cost and makes the production costlier and less economical. So if corrosion rate of steel (it is an alloy having nearly 98–99% of iron) is not significant, its usage will remain economical. However, if we inoculate this media by adding some bacteria, corrosion of metal may become significantly high, as measured from corrosion rate or pitting/crevice corrosion, etc., so as to turn the metal uneconomical to use. This corrosion is termed 'microbial induced corrosion' or MIC. So to observe and explore MIC, it is necessary to expose metal to an inoculated media which should be such that metal experiences corrosion mainly due to bacteria. Obviously, this media should be such that (i) it corrodes metal only minimally due to its chemical constituent and (ii) the bacteria, in question, should survive and grow , in this media, with time.

- Conditions for Survival and Growth of Bacteria

Now, what should be the constituents of this media and what should be its pH and temperature at which the experiment for estimating MIC be conducted. Should this media be aerobic or anaerobic and/or whether it will require presence of some gas at a certain pressure in order that bacteria survive in it, etc. Sometimes, particularly, when the experiment is to be performed under conditions which simulate some industrial process, e.g., MIC of material of pipelines which is meant for conveying liquid media, filters, and washers. In such cases, the material experiences corrosion not just due to constituents of liquid media alongside bacteria but also the erosion effect due to its movement. It may then become necessary to impart turbulence in the media with the help of a rotor so that experimental conditions simulate those of the industrial process. Another case could be to do the experiment under high pressure/vacuum, low temperature conditions. Example of such cases could be if one intends to simulate the experimental conditions with those which material of construction experiences as a part of digester/pressure vessel, or as a part of low temperature enclosure or a cryogenic system. In many such cases, the bacteria may simply not survive so question of material experiencing MIC does not arise. So one has to look at the conditions (Table 4.1) under which different bacteria survive [1], which in turn will help in finalizing the composition of the test media for investigating MIC in the presence of a particular bacteria.

Reactions as experienced by electron acceptors (§ in Table 4.1) are (1.5b), (2.1), (2.2), (2.4) and those given below:

$$CO_2 + 8H^+ + 8e^- \rightarrow CH_4 + 2H_2O \tag{4.1}$$

$$NO_2^- + 4H^+ + 2e^- \rightarrow NH_4^+ + O_2 \tag{4.2}$$

$$2NO_3^- + 10H^+ + 8e^- \rightarrow N_2O + 5H_2O \tag{4.3}$$

$$NO_3^- + 2H^+ + 2e^- \rightarrow NO_2^- + H_2O \tag{4.4}$$

$$S^0 + 2e^- \rightarrow S^{2-} \tag{4.5}$$

$$S^0 + H^+ + 2e^- \rightarrow HS^- \tag{4.6}$$

These reactions help in understanding the thermodynamic considerations in order to know whether, as a result of these reactions, metal will experience corrosion, will remain immune to corrosion or will passivate. This in turn helps to understand the bioenergetics of MIC and its mechanism.

Table 4.2 gives a list of electron acceptors alongside the respective organisms which get energy for survival as a consequence of these reactions alongside oxidation reaction. These electron acceptors undergo one or more of the reactions vide Eqs. 1.5b, 2.1, 2.2, 2.4, and 4.1–4.6 during MIC.

Though organisms are normally found at temperatures ranging in thirties (30 °C or so) and neutral pH, but some bacteria do exist under conditions which would be considered impossible for the existence of human beings. Range of conditions under which microbes are observed to exist are given in Table 4.3

Thus, with the help of Tables 4.1 and 4.2, one can decide about electron acceptors which will be responsible for reduction reaction and hence the rate of corrosion of electron donor, i.e., a metallic material, source of carbon meant for growth of bacteria so that bacteria do not die during the experiment. The first two components are also responsible for providing energy (Chap. 2) to bacteria. Table 4.3 helps one in knowing the extreme conditions of composition of media which is to be inoculated with the bacteria to be tested for MIC.

Second condition is that the experimental set-up should provide a surface where biofilm due to bacteria may grow. This surface could be that of a metal, e.g., iron, copper, steel, stainless steel, and brass, etc., as industrial materials which experience microbial corrosion during the experiment. These materials also become source of energy (Chap. 2) for the growth of bacteria which is so essential for observing MIC.

- Isolation of Bacteria from Natural Resources and Its Identification

For investigating MIC due to a given bacteria, the same should be either obtained from standard source or it may have to be isolated from natural sources or from an industrial machinery which may have suffered MIC due to some bacteria. There may

Table 4.1 Requirements for survival and growth of microorganisms

Prerequisite	Provided by	Kind of Growth
Energy source	Light	Phototrophic
	Chemical substance	Chemotrophic
Carbon source	CO_2	Autotrophic
	Organic substances	Heterotrophic
Electron donor (which is oxidized—reducing agent)	Inorganic substances, e.g., metal*	Lithotrophic
	Organic substances**	Organotrophic
Electron acceptor § (which is reduced—oxidizing agent)	Oxygen	Aerobic
	NO_2^-, NO_3^-	Anoxic
	SO_4^{2-}, CO_2	Anaerobic

*Essential for cases of microbial induced corrosion (these may be steel, stainless steel, Cu, brass, etc.)
**Could be constituent of growth media in dissolved state, e.g., lactates in reaction (2.3)

Table 4.2 List of electron acceptors alongside respiration and organisms

S. No.	Electron acceptor	Product	Name of respiration and organisms
Aerobic respiration			
1.	O_2	H_2O	Oxygen respiration Aerobic organisms in presence of O_2 *Pseudomonas aeruginosa, E. Coli*
Anaerobic respiration			
1.	NO_3^-	NO_2^-, N_2O, N_2	Nitrate respiration *Pseudomonas stutzeri, Paracoccus denitrificans*
2.	SO_4^{2-}	S^{2-} (HS- for lower pH)	Sulfate respiration, obligate anaerobe bacteria *Desulfovibrio desulfuricans, desulfotomaculum ruminis*
3.	S	S^{2-} (HS- for lower pH)	Sulfur respiration *Desulfuromonas acetoxidans*
4.	CO_2	Lactate Methane	Carbonate respiration acetogenic bacteria, *Clostridium acetium* Carbonate respiration Methanogenic bacteria
5.	Fe^{3+}	Fe^{2+}	Iron respiration

Table 4.3 Range of known conditions for observation of microbial life

Parameters	Range
Temperature	-5 °C (salt solutions) to >120 °C (hot vent at the bottom of sea)
pH	0 (Thiobacillus thiooxidans) to 13 (Plectonema nostocorum, in mineral rich soda lake)
Redox potential	Entire range of water stability -450 mV (methanogenic bacteria) to $+850$ mV (iron bacteria)
Pressure	Up to 1.000 bar (*barophiles* survive with optimized reproduction at high pressure in deep sea. *Pyrococcus yayanosil* can survive up to 1500 bar)
Salinity	*Burkholderia cepacia* in ultrapure water (salinity ~10–200 ppm) to 'halophilic bacteria' almost saturated water (in Dead sea, salt content is 280 g l^{-1} or 280,000 ppm)
Nutrient concentration	10 µg/l (drinking and purified water)—carbon sources

be two different types of samples (i) solid and (ii) liquid. In the first case, representative samples should be taken from undisturbed machinery part on or adjacent to metal surface. Soil, corrosion products, deposits and microbial slimes, on the metal surface of machinery part, could be the suitable materials for sampling. These may be scrapped from metal surface using a tool which avoids contamination and collected in a vessel tightly closed and cool during transport. Care should be taken to avoid any transformation or change in the bacteria, e.g., on exposure to oxygen, due to leakage in packing, some anaerobes are irreversibly inhibited. In these cases, the quantity of sample collected should be larger to ensure anaerobic conditions within some part of sample at-least. In case of liquid sample, the source of bacteria which could be river, any water body, or some process liquor from some plant machinery, etc., is mixed in a media suitable for growth of bacteria which we intend to isolate. This is followed by the segregated bacteria incubated at room temperature for several hours for increasing its concentration. The culture then is serially diluted in a suitable media and spread over a suitable surface, e.g., winogradesky agar medium. After several hours of exposure, colonies of the desired bacteria appear. At this stage, pure culture of the bacterial isolate is obtained by streak plate method. The bacterial isolate are preliminary identified by different biochemical tests, after which they are identified using 16SrRNA sequencing by microbial cultural collection. Due to lesser concentration of bacteria in liquid samples, the amount of sample collected in these cases should be larger than in case of solid samples.

As an example, method for isolating iron-oxidizing bacteria from river water is described here (4.2). For its isolation, water samples were collected from river in sterile container. Steel coupons were immersed in this water for 15 days which resulted in rusting of steel coupon. This rust was scrapped by sterile spatula and used to inoculate winogradesky broth medium (WBM). The inoculated medium was incubated at 37 °C for 24–48 h. The incubation of bacteria leads to change of media color from green to rusty brown indicating the presence of iron-oxidizing bacteria due to oxidation of ferrous to ferric iron. Afterward, the culture is serially diluted in WBM and spread over winogradesky agar medium for 24–48 h when rusty brown colored colonies of the bacteria appear. Pure culture of bacterial isolates was then obtained by streak plate method. The isolated bacteria were then identified by different biochemical test and finally by 16SrRNA test.

• Growth of Bacteria

Growth of bacteria is measured by most probable number (MPN) method. It is based on the principle that when the material containing bacteria is cultured, every viable bacterium produces a visible colony on the agar medium. The number of colonies is the same as the number of organisms present in the medium. Growth of bacteria is evaluated after every fixed interval of time. One can then view it on a growth curve plotted between bacterial concentration and time (Fig. 4.1a) to estimate bacteria's lag phase (0–12 h), exponential phase (12–48 h), and stationary phase (48–84 h) followed by death phase. These parameters, e.g., bacteria concentration during 'stationary phase' may be used in determining the composition of test solution for corrosion tests, etc.

• Biofilm Characterization

This is done in terms of concentration of solid components of biofilm namely termed as 'extracellular polymeric substances' (EPS). It is well known that biofilm is mainly water (~95%), rest is solid components. The EPS of a biofilm consists of carbohydrates, proteins, uronic acid, and lipids. For estimating these components, biofilm is

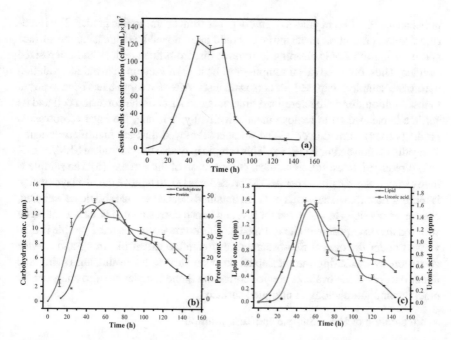

Fig. 4.1 a Growth curve of a bacteria in terms of sessile cell concentration versus time. **b, c** Biofilm characterization in terms of concentration of EPS of biofilm [2]

removed from the surface of an exposed metal coupon which has been exposed to an inoculated solution. The removed biofilm is suspended in distilled water. The solution, thus obtained, is centrifuged at a given RPM for several minutes. The supernatant collected after the centrifugation is filtered using 0.22 micron cellulose acetate filter for removing cells. Cell-free filtrate is used for estimating the amount of components of biofilm by spectrophotometric methods using standard procedures. Figure 4.1b, c shows the concentration of all the four components of biofilm of iron-oxidizing bacteria.

4.2 Experimental Set-up—Corrosion and Related Aspects

- Metal Surface Preparation

As discussed in Chap. 1, metal surface preparation is a crucial part of corrosion tests, since presence of cracks, scratches, or non-uniformity of composition affects the behavior of samples against corrosion. So to avoid these dependences, material chosen should be crack free, it should be heat treated, in vacuum, appropriately to remove non-uniformities, and to remove scratches it should be abraded and polished using different grades of emery paper from coarse to fine grades. The impurities/grease present on metal surface should be removed by degreasing the surface and by using ultrasonic cleaning to remove dirt from fine crevices and microsized cavities. Thus, obtained metal sample must be kept in vacuum to avoid its oxidation before the sample is exposed for corrosion tests. After the sample has been exposed to test solution, for a predetermined time, it is taken out of the corrosion cell and its surface is cleaned (i) to remove mechanically, by brushing, the layer of corrosion products (rust) formed over it. The products removed thus are identified by using X-ray diffraction and/or FTIR, etc. This sample may also be undertaken SEM/EDAX test, if required, to estimate chemical composition of the surface. (ii) The sample is then cleaned chemically, as per ASTM procedures [3], to remove the rust layer which is more adherent to metal surface. Thus, obtained cleaned sample should be weighed, as soon as possible, to know weight loss and hence corrosion rate (Chap. 1) of metal suffered due to exposure in the test media. This corroded and cleaned sample is then viewed under the optical microscope to estimate the extent of pitting and crevice corrosion by measuring the pit depths on open surface for evaluating pitting and under the crevices to evaluate crevice corrosion. Same samples are also analyzed for cracking and morphology of corroded surface.

• Preparation of Metal Sample and Test Solution

The metal coupons are abraded and polished for surface preparation as described previously. These coupons are placed under vacuum in dessicator. They are taken out, just before putting for corrosion test, for degreasing in acetone and sterilizing by exposing to 70% ethanol for several hours followed by drying under ultraviolet light in a stream of warm air. The test solution is prepared of a predefined composition and is inoculated with bacteria. The pretreated metal samples, as described above, are put into inoculated test solution for a given time after which they are taken out for analyzing various aspects related to MIC experienced by them.

• Corrosion Test

Immersion Test: There can be two different situations—one when single or duplicate samples of the same material is/are to be tested and other where multiple samples (different materials) are to be tested. In the former case, the metal coupon, in duplicate, is placed as in Fig. 4.2a with serrated washers meant for studying crevice corrosion or corrosion under deposit alongside pitting and uniform corrosion. The coupons are kept electrically separated from other metallic components, e.g., nuts and bolts, etc., by spacers or serrated washers. All the spacers/washers are of plastic.

Figure 4.2b is a case where more than one type of metals is to be tested simultaneously. In this case, duplicate coupon of each metal is fixed in a rack separated by spacers/serrated washers. Here too, all the metal coupons are kept electrically isolated from all metal components of the rack. In either case, the coupons are put in the test media for a specific time (normally 3–6 months) for measuring the extent and type of corrosion. An idea of estimating the time of exposure and amount of test media is suggested in Chap. 1.

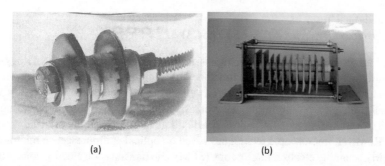

(a)   (b)

Fig. 4.2 **a** Duplicate metal coupons with serrated washers, **b** Duplicate coupons of more than one type of metals meant for testing metals simultaneously

Electrochemical Test: This experiment is performed in a 5 neck cell (Fig. 1.18). The metal sample, to be tested, is put in this cell in a specific holder depending upon whether sample is flat (Fig. 4.3a) or cylindrical (Fig. 4.3b).

Other details regarding corrosion cell, experimental set-up, etc., are given in Chap. 1. From these experiments, one can know quickly within at the most 2–3 h (short term) about the corrosivity of inoculated solution and the resistance of the test metal against uniform corrosion, pitting, and crevice corrosion. These experiments also help in understanding the mechanism of corrosion in various cases through the electrochemical aspects of corrosion.

Although both type of tests, immersion tests and electrochemical, give similar information regarding corrosion but they are complimentary to each other. Selection of a material for fabricating industrial machinery is done from some of the possible candidate materials which have enough resistance against corrosion in a given media. These materials are shortlisted on the basis of their performance against corrosion in the media, which the fabricating material needs to be in close contact throughout, using the electrochemical techniques. Most appropriate material from these short-listed materials, which may be used for fabricating the machinery, is selected on the basis of their corrosion performance as judged on the basis of immersion test. This is so because the selected material has to perform without unexpected or sudden failure due to corrosion considering lifetime of several years. For this purpose, immersion test is more appropriate since it gives the picture of corrosion performance on long term basis.

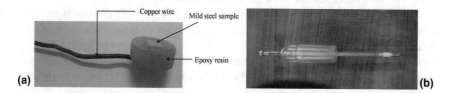

(a)   (b)

Fig. 4.3 Sample holder for **a** flat specimen and **b** cylindrical specimen

- Analytical Tests
  (*Spectrometry, XRD, FTIR, SEM, ESCA*)

Wide range of microscopic techniques is used in the field of microbiology and so in the case of studying MIC too. The conventional microscopy techniques are not capable of visualizing the bacterial cell surfaces or the events involved in MIC in real time, e.g., pitting initiation, synthesis of extracellular polymeric substances (EPS), or transition from initiation of MIC to its inhibition on the affected surface, etc. However, microscopic techniques based on optical epifluorescence microscopy (OEM), scanning electron microscopy (SEM), and transmission electron microscopy (TEM) provide information about elemental composition while 2D TEM data about data and subjects. However, one of these techniques (OEM) does not have sufficient resolution while other two (SEM and TEM) have two shortcomings—first they require sample preparation so cannot be used on-site, second the sample distortion caused by the dehydration process. Atomic force microscopy (AFM), environmental SEM (ESEM), and confocal laser scanning microscopy (CLSM) are used for observation of biofilm in real time without transformation in the sample structure. AFM, which has a resolution of nanometer, can give one an idea about a cell and its associated EPS on the surface of a metal. CLSM has shown that 75–95% of the volume of bacterial biofilms is occupied by the matrix, and cells may be concentrated in only 5–25% of the lower or upper layers. By using ESEM, one can record high resolution images for detailed observation of biofilm without dehydrating. Using CLSM and AFM, one can examine three-dimensionally hydrated and living biofilm in real time.

For analysis of corrosion products, electrochemical mechanism of MIC in various cases, etc., one needs to characterize biofilm formed on metal surface, elemental composition, corrosion products formed, etc. While a discussion on biofilm characterization has been given earlier, a brief idea about SEM/EDS, ESCA, X-Ray diffraction, and FTIR is given here. While SEM is used for knowing morphology of biofilm formed over metal surface, a quantitative idea about elements present in the biofilm can be had from energy-dispersive X-ray spectroscopy (EDS), an attachment for which is available in SEM/EDS system. The principle involved in this technique is the interaction of electron beams with various atoms which could be part of biofilm, cells, corrosion products present on the top of corroded metal surface. The exchange of electron energy with atoms emits X-rays of different wavelengths characteristic of the atomic structure. The result is displayed in terms of intensity vs. energy of X-rays emitted from various atoms. The system is programmed to show the result in terms of atomic composition vs. their respective amount.

X-ray diffraction technique is a standard technique to identify corrosion products. In this technique, X-ray diffractometer, having normally Cu target for emitting $CuK\alpha$ radiation, is used. The technique works on principle of diffraction of X-rays (Bragg's Law) which is characteristic of structures of various inorganic compounds, e.g., corrosion products namely α, β, γ-FeOOH, Fe_2O_3, iron hydroxides, iron sulfides, etc. The X-ray diffractogram obtained is a variation of intensity of X-rays diffracted at various angles, which is characteristic of structure of different compounds. Data from ASTM data cards [4] help in identifying various possible corrosion products formed as a result of MIC experienced by metal sample. This information is further verified by using Fourier transform infrared spectroscopy (FTIR). The technique also gives information on various chemical bonds like C-O,

C–H, P=O, etc., functional groups. These details characterize the biofilm and its EPS. Information on surface chemical analysis tells about corrosion products and microbiological deposits present within a few nanometer on the surface. For this purpose, one uses a technique which employs electrons as probe because they have less energy so they cannot penetrate much hence interact at depths only within few nanometers of the surface. Electron spectroscopy for chemical analysis (ESCA) is used for getting this information. Auger electron spectroscopy (AES) allows mapping of corrosion products across a metal surface that has undergone localized corrosion, e.g., pitting, stress cracking, etc. This technique has been used in investigating biocorrosion in condenser tubes. Fundamentals and details about these techniques can be sought from Internet facility.

4.3 Corrosion-Causing Bacteria and Strategies to Control MIC

In this section, brief description is given about various studies on metal corrosion due to presence of bacteria. MIC cases are observed in cases where some of the microorganisms are found to dramatically accelerate corrosion kinetics particularly where liquid media is having practically no oxygen and pH is around 6 or so. Just to remind, relaxing the two later conditions, e.g., presence of oxygen and lowering of pH enhances corrosion even in absence of bacteria. This section is discussed in two parts—corrosion enhancing bacteria, i.e., cases of MIC and corrosion inhibiting bacteria. The later section is important from the view point of green technology approach for mitigation/inhibition of corrosion as well as MIC.

- Corrosion-causing Bacteria
  Sulfate-Reducing Bacteria (SRB)

Sulfate-reducing bacteria (SRB) is a group of anaerobes that reduce oxidized sulfur compounds, e.g., sulfate, thiosulfate, sulfides as well as sulfur to H_2S. This reaction alongside the metal oxidation reaction helps giving the bacteria the required energy for their growth. MIC due to SRB's is a very widely studied topic due to (i) their wide prevalence in various industrial systems and so effect their economy of production and (ii) their capability of enhancing the reduction reaction rate significantly which in turn makes them highly corrosive in nature. SRB are observed in process media of very many industries and therefore affect, severely, many of their machinery components. Few of the examples are given here. Thermophilic SRB have been observed to be responsible for severe intergranular pitting of 304L stainless steel condenser tubes in geothermal electrical power plant operating at >100 °C. In another example, microbiological activity and chloride concentrated under scale deposits were blamed for the worm hole pitting of carbon steel piping used to transport a slurry of magnesium hydroxide and alumina at pH 10.5. SRB are also found responsible for external corrosion under the disbonded coating on a cast iron pipeline buried under the anoxic (without oxygen) conditions and sulfate-containing soil. Although SRB's are anaerobic but some genera tolerate oxygen and some sp. of SRB are able to grow at low oxygen concentration. These bacteria corrode various installations of steel through MIC in many industries, e.g., oil and gas, shipping, etc. Biogenic sulfide production (by SRB's) also leads to severe economic loss due to reservoir souring which results in failure of stainless steel equipment through sulfide stress and hydrogen-induced cracking and other corrosion effects. The deterioration of steel structures in freshwater and marine environments is crucially dependent on the MIC due to SRB and so attract special attention.

In petrochemical plants, pipelines and fittings are considered to be strategic components, especially when it comes to sustainable and safe operation. Among different fittings, elbows, mostly with angles of 90° or 45°, are the most used ones. Elbows can experience severe damage in the case of dramatic changes in the flow pattern. Materials of these elbows are carbon steels for transporting petroleum products and water. SRB under such condition cause severe biodegradation of the external surface of buried pipelines. As far as SRB-induced biocorrosion of the interior of pipes is concerned, the low content of oxygen, together with stagnant liquid inside the pipelines provide desirable conditions for SRB to grow. Additionally, SRB can appear under deposits of soil, water, hydrocarbons, chemicals, etc. It is well known that kinetics of corrosion in steels under the influence of microorganisms, including SRB, can be up to ten times faster than the kinetics in the absence of microorganisms. Presence of SRB's in seawater is also of concern for maritime vehicles, e.g., ballast tank, ship, etc. A ballast tank is a compartment within a boat/ship/submarine, etc., that holds water to adjust the marine vessel's draft, buoyancy, etc., under different operating conditions. These could be like weight reduction, due to fuel consumption or stormy weather, and can be balanced by increasing vessel's stability. They can be filled when cargo is being offloaded or emptied while loading the cargo. Presence of water also includes presence of different bacteria including SRB's. These ballast tanks are prone to MIC as they are continuously filled or emptied leading to high degree of humidity (95%) level which causes high corrosion rates of steel plates peaking around 2–5 mm/year.

To quantify the extent of corrosion and to find out factors responsible for higher degree of corrosion due to SRB, many studies have been conducted [5–7]. Accordingly, iron corrosion in anoxic environment, to ensure that corrosion is due to bacteria mainly, is influenced by microorganisms, particularly, sulfate-reducing bacteria. The corrosion process is characterized by formation of black crusts and metal pitting. The electrochemical reactions involved in this process are (1.1), (1.6), (2.1), (2.5), and those given below:

$$H_2O \rightarrow H^+ \; OH^- \qquad\qquad (4.7)$$

$$OH^- + HCO_3^- \rightarrow CO_3^{2-} + H_2O \tag{4.8}$$

$$Fe^{2+} + CO_3^{2-} \rightarrow FeCO3 \tag{4.9}$$

$$4Fe + SO_4^{2-} + 8H^+ \rightarrow FeS + 3Fe^{2+} + 4H_2O \tag{4.10}$$

Cathodic hydrogen formed on metal surface can promote growth of SRB through available energy form redox reactions (1.1 and 1.6). The atomic hydrogen escapes the metal surface by combining to form molecular hydrogen or it enters into the metal matrix as atomic hydrogen. FeS and FeCO$_3$ are black hard minerals deposited on corroding metal. These minerals exhibit electrical conductivity. Iron sulfide together with iron forms a galvanic couple with former acting as cathode and later as anode. This results in corrosion of metal. It was also observed that higher the amount of formation of iron sulfide higher was the corrosion rate of metal. This is in line with basic principle of corrosion that higher cathodic area in a galvanic couple leads to more corrosion of the anodic material. In the experiment performed, metallic iron was exposed to marine SRB, 72% of iron coupons degraded in 5 months corresponding to a corrosion rate of 0.7 mm/year. Iron coupons in such case were covered with black crust. However, corrosion rate in aquatic case, with same chemistry but without bacteria, this rate was much less. It was also observed in another experiment that sulfide generated out of bacterial activity was highly corrosive while chemically generated sulfide does not have same degree of corrosivity. Thus, MIC observed due to SRB is associated with production of biogenic sulfide. Studies done on the two sulfides, biogenic generated and abiotic produced, demonstrated different structure and composition. It appears that one predominant mechanism may not be existing in case of SRB-induced corrosion.

It is quite likely that more than one species may exist of the same bacteria. It will be of interest to see if species of the same bacteria influence corrosion on a given metal to same extent. SRB's are one of many such cases, e.g., one observes *Desulfovibrio desulfuricans* and *Desulfomaculum nigrificans* as the two species of SRB's. They were tested for their degree of corrosivity on stainless steels in nutrient-rich Baar's medium. Based on electrochemical tests and immersion test [8], it was observed in case of *Desulfovibrio desulfuricans* that extent of uniform corrosion and pitting was more in case of inoculated Baar's media than the control media. Further, tendency of stainless steels to experience localized corrosion was observed to increase with increased cubation time in the inoculated solution perhaps due to higher bacterial concentration. Observance of change of pH during test to higher values indicates it due to the formation of H$_2$S and thus supports cathodic depolarization theory of SRB-induced corrosion. In another study, MIC on the same stainless steels were tested in the same media (nutrient-rich Baar's media) but inoculated with another SRB sp. *Desulfotomaculun nigrificans* [9]. The degree of corrosion uniform as well as localized in this case was observed higher, on respective stainless steels, than in case of SRB sp. *Desulfovibrio desulfuricans*. Higher degree of corrosivity can be attributed to the presence of higher amount of protein in the harvested EPS of the former species of SRB.

Since different industries are using different liquid media, based on their raw material and product, bacteria of the same genera found in them are surviving in diversified environment. As an example, SRB's are observed in oil and petrochemical industry, paper mills, seawater where maritime vehicles move, etc. The process liquors in the two industries and seawater have different chemistry. Now, do same bacteria present in different media affect, through corrosion, the materials of construction/fabrication of machinery/equipment similarly or in different ways. In another study [10], it was, therefore, attempted to answer this question by conducting MIC tests on stainless steels in Baar's media, sodium chloride solution and artificial seawater inoculated with SRB sp. *Desulfotomaculun nigrificans*. Maximum corrosion attack was observed in case of exposure to seawater followed by sodium chloride solution and minimum in case of Baar's medium. This trend appears to correlate with the highest amount of all the EPS components, i.e., carbohydrates, protein, uronic acid and lipids of the biofilm formed on stainless steels exposed to seawater. Respective amount of these components is lower in case of exposure in sodium chloride solution and lowest in case of Barr's media. It is known that EPS have tendency to bind metal ions which lowers down their concentration in the electrolyte which in turn enhances oxidation hence corrosion of metal.

Iron-Oxidizing Bacteria (IOB)

Iron-oxidizing bacteria are aerobic microorganisms, belonging to a large and diverse group. Most common iron-oxidizing bacteria are the species *Thiobacillus ferrooxidans* and the genera *Crenothrix, Gallionella, Leptothrix, and Spherotillus.* They are widely found in water from rivers, lakes, and oil production plants. As such, many industries namely power generation, water cooling system, pulp and paper, oil production, and petrochemical, etc., are affected by corrosion due to this bacteria. Though their extent of corrosivity is not as detrimental as that of SRB's still their corrosion effect is significant enough not to be overlooked. In some areas, particularly, in oil exploration where SRB are also observed alongside IOB, the degree of corrosivity is much more than that observed in case IOB or SRB alone. The IOB's get energy necessary for their metabolism from oxidation of Fe^{2+} to Fe^{3+} iron. Consequently, there is the formation of iron hydroxides that generally form insoluble precipitate on the surfaces, promoting regions with different oxygen levels and formation of rust. The accumulation of rust products or inorganic fouling leads to blockages of pipelines. Thickness of the deposited layer is observed to range from 1 to 6 cm. Iron bacteria, mainly *Sphaerotilus genu*, were isolated from these rust sediments. Severe pitting attack (pit depth up to 10 mm) has been detected beneath these deposits. Various cases of corrosion damage caused by iron-oxidizing bacteria have been observed in chemical and water industries. The majority of these cases occurred at oil refinery plants with carbon steel equipment such as: heat exchangers (tubes, lids, tube-sheets, connection pipes), extinguisher pipelines, and other parts of the water system.

Another industrial application, important from standpoint of MIC due to *P. aeruginosa*, is the corrosion of Ni-Co alloy by IOB's since these alloys are used for nickel electroplating of carbon steel plates which in turn provides strength, toughness, and corrosion/wear resistance to the substrate. Presence of IOB leads to accumulation of biofilm on plated steel surface causing rupture of nickel coating followed by pitting, and crevice (under deposit) corrosion of the underlying steel plate.

Leptothrix bacteria is filamentous iron bacteria and are 'omnipresent' in carbon steel and iron distribution system pipelines. These bacteria are commonly reported in deposits associated with tuberculation. It was reported that Leptothrix sp. would grow autotrophically utilizing the energy liberated upon the oxidation of ferrous to ferric ions. These are the most common iron storing ensheathed bacteria apparently occurring in slow running ferrous iron-containing waters. The sheaths of *Leptothrix* sp. assist in the formation of a membrane that is relatively impervious to oxygen, and in the process decrease the quantity of oxygen in the tubercle vicinity, thus establishing a micro-electrochemical cell. With increasing thickness, the inside of the tubercle becomes more anaerobic, the difference in the potential between inside and outside the tubercle increases, and corrosion gets accelerated.

Pseudomonas aeruginosa is an iron-oxidizing bacteria predominant in marine environments. It is, therefore, expected to affect the carbon steel parts of marine vehicles. With the possibility of replacing these parts by those formed of stainless steel, a study [11] was performed to investigate the corrosivity of these bacteria on stainless steel 304 in nutrient-rich simulated seawater (NRSS). The metal samples were tested in both sterile and inoculated NRSS medium. Pitting was observed in both the cases but the severity of pitting on stainless steel sample was higher when exposed in inoculated than the sterile media. Obviously, even the stainless steel parts are likely to experience limited life before failing due to pitting.

Several other studies on IOB have appeared in the literature. Here is a brief idea given about them. *Sphaerotilus* sp., an IOB, has been reported to initiate pitting in 3% NaCl solution. *Acidithiobacillus ferrooxidans* has been reported to alter the pH of solution and produces dense and thick layer of Fe_2O_3 on the surface of steel that leads to enhanced corrosion. Another study reported production of thick oxide layer by IOB, on stainless steel, thus creating a suitable environment for anaerobic bacterial growth, e.g., SRB. The two bacteria synergistically result in higher corrosion rate as compared to IOB alone. Pseudomonas sp., an IOB, is observed to inhibit the growth of planktonic SRB but promoted the sessile SRB in mixed culture and accelerate corrosion of carbon steel.

In order to study the dependence of corrosion due to IOB on biofilm, corrosion products, etc., formed on steel with the aim of understanding the mechanism of MIC, investigations have been performed recently [12] on corrosion of mild steel in winogradesky media inoculated with IOB sp. *Pseudomons* which was isolated from river water. Immersion test and electrochemical polarization test together with characterization of biofilm (by optical spectrometric method) and identification of corrosion products (by X-ray diffraction and FTIR method) produced on exposed steel samples formed the basis of analysis. It was concluded that (i) presence of bacteria enhances uniform and localized corrosion and (ii) increased corrosion can be attributed to presence of wider cracks on the biofilm, carbohydrates of the biofilm, and presence of iron sulfite as the corrosion product. Based on these findings, a mechanism of corrosion has been proposed. Further tests have been carried with the aim to test the dependence of MIC on (i) species of *Pseudomonas aurginosa* and (ii) host media winogradesky and nutrient broth. Among the species, it is observed that sp. DASEWM2 imparts higher degree of corrosion attack on steel than sp. DASEWM1. This can be attributed to higher amount of carbohydrate as a major component of EPS in the biofilm of bacteria. Among the host media, inoculated winogradesky media appears more corrosive than inoculated NB media. Lesser corrosivity of inoculated nutrient broth media could be attributed to presence of vivianite in the corrosion products which are protective in nature whereas higher corrosivity of inoculated winogradesky media appears to be due to absence of vivianite and presence of sulfite, which is corrosive type, in the corrosion products formed in this media. It is thus seen that extent of corrosion attack, on steel, due to a particular bacteria is dependent upon its species as well as the host media which is being inoculated.

Metal-Reducing Bacteria

There are some bacteria, e.g., from the genera of *Shewanella* and *Pseudomonas* which enhance corrosion of steel and stainless steel by inducing reactions which dissolve corrosion resistant oxide films on the metal surface like Cr_2O_3, molybdenum oxide on stainless steel, etc. Both these oxides passivate stainless steel so on their dissolution the passivation is lost or may get replaced by less stable reduced metal films. This results in increased corrosion attack on metal. Such reactions are quite wide spread in nature and may have lot of industrial implications in enhancing corrosion of machinery. However, they have not been investigated seriously.

Acid-Producing Bacteria

Some bacteria produce large quantities of inorganic/organic acids as a result of metabolic reactions. Thus, inorganic acids like nitric and nitrous acid, sulfuric and sulfurous acid, and carbonic acid are produced by mainly bacteria of the genera *Thiobacillus*. Thiobacilli can tolerate extremely acidic conditions and can grow at pH of 1. Nitric/Nitrous acid are produced by bacteria of ammonia- and nitrite-oxidizing group. The problem of corrosion arises due to the fact that salts resulting from sulfuric and nitric acid are water soluble hence cannot provide any protection. Moreover, due to the lowering of pH protective deposits, e.g., $CaCO_3$ are dissolved.

Methane Producers

Methanogens, bacteria which produce methane, e.g., *M. deltae, M. thermolithotrophicus, M. berkeri*, enhance corrosion of iron, steel through MIC. The MIC occurs through cathodic depolarization of CO_2 present in the media for growing the methanogens. Electrochemical reactions in the process are (1.1) as oxidation and (4.1) as reduction reactions. The above reactions are thermodynamically favorable since $\Delta G° = -136$ kJ/mol of methane at pH 7.0 and 37 °C. From the reactions, it is clear that the reactions will be more favorable at lower pH due to higher concentration of H^+ ions in the media. Same conclusion one draws when $\Delta G°$ is calculated for lower pH vales. Thus, $\Delta G° = -182$ kJ/mole and -228 kJ/mole at pH 6 and 5, respectively. To ensure that increase in corrosion at lower pH is due to biological reaction and not because of chemical reaction due to more acidic conditions, experiments were performed [13] to measure corrosion rate of iron coupons in media, with and without bacteria, at pH ranging from 5.2 to 7.5. At every pH, corrosion rate of coupons was higher in inoculated media, although in case of abiotic media, reduced pH was also observed to accelerate corrosion always, but to lesser extent.

Ammonia Production

Bacillus sp. *AG1* and *AG3* are the bacteria which corrode Cu, Cu-Ni alloys through ammonia production. They are resistant to copper toxicity and were capable of attacking mirror polished cupronickel (Cu-Ni alloy) coupons in the medium when ammonia was released through biogenic ammonical action. Corrosion tests [14] were conducted in nitrogen-free water without bacteria, from a source from which the *Bacillus* sp. bacteria were isolated and identified, taking it as control. Two test solutions were prepared from this nitrogen free water, for the study, one inoculated with AG1 and another with AG3. Corrosion rate obtained from weight loss measurements were 0.008 mm/year for control, 0.023 mm/year, and 0.030 mm/year in test solutions with bacteria AG1 and AG3, respectively. Since these bacteria produces ammonia, reduction reaction (1.5a) and following electrochemical reactions are considered to be responsible for enhanced corrosion rate as well as observance of intergranular corrosion:

$$NH_3 + H_2O \rightarrow NH_4^+ + OH^- \qquad (4.11)$$

$$Cu + 2OH^- \rightarrow Cu(OH)2 + 2e^- \text{ (oxidation)} \qquad (4.12)$$

$$Cu(OH)_2 + 2NH_3 + 2NH_4^+ \rightarrow \left[Cu(NH_3)_4\right]^{2+} + 2H_2O \qquad (4.13)$$

The enhancement of anodic current in the polarization study can be explained by the fact that the corrosion on cupronickel was due to the oxidation of Cu. Formation of [Cu $(NH_3)_4$]$^{2+}$ ions, which is consequence of Cu oxidation, results in depletion of Cu in intergranular region leading to Cu rich grains acting as cathode and Cu depleted region in intergranular region as anode. Hence, the intergranular attack can be linked to the over saturation of the [Cu $(NH_3)_4$]$^{2+}$ ions. Further, these ions convert to CuO as per the following reaction

$$\left[Cu(NH_3)_4\right]^{2+} + H_2O \rightarrow CuO + 2NH_3 + 2NH_4^+ \qquad (4.14)$$

The formation of CuO at the metal surface, noticed in X-ray diffractogram, is a critical step needed for the occurrence of stress corrosion cracking (SCC).

- Corrosion Inhibiting Bacteria

As we know that any surface, be it metal, plastic, ceramic, or whatever, on exposure to a biotic environment which could be a liquid, moist vapor, etc., gets covered with biofilm characteristic of the bacteria it has. This biofilm may be loose, compact, porous, or it may be conducive to diffusion of ions/chemicals or passage of electrons. Further, its EPS constituents may react with chemicals of the biotic media along with the metal constituents so as to produce sometimes corrosive environment and/or chemicals or may change the chemistry of the environment to a lesser corrosive one. While the former case is responsible for MIC but it is the latter case that acts to inhibit the MIC leading to a phenomenon known as microbial induced corrosion inhibition (MICI). Such bacteria are termed as corrosion (read MIC) inhibiting bacteria.

Corrosion control using beneficial bacteria through their biofilms, in principle, can be achieved by several possible mechanisms: (1) removal of corrosive agents (such as oxygen) by bacterial physiological activities (e.g., aerobic respiration), since oxygen in aerobic media is an important oxidizer of metals by undergoing reduction. There are other oxidizers also quite commonly observed, e.g., Cl_2, hypochlorite, etc., (2) growth inhibition of corrosion-causing bacteria by antimicrobials generated within biofilms, e.g., sulfate-reducing bacteria (SRB) corrosion inhibition by gramicidin-S-producing Bacillus brevis biofilm, (3) generation of protective layer by biofilms, e.g., *Bacillus licheniformis* biofilm produces on aluminum surface a sticky protective layer of γ-poly glutamate. (4) Latest in the series is by the use of extracellular polymeric substances (EPS) produced by bacteria for inhibition [15]. In this method, various strains of bacteria, e.g., SRB, Pseudomonas, etc., are cultivated and their EPS is harvested. These EPS are then tested for their intrinsic corrosiveness toward different metal substrates by adding droplets of EPS from various bacteria to the control, e.g., water, etc.

Microbial Inhibition in Aerobic Condition Due to Removal Of Oxygen

Let us consider corrosion of SAE 1018 carbon steel in sterile aerobic media having near neutral pH. In this condition, corrosion of steel proceeds though oxygen reduction as cathodic reaction. So if we somehow remove or decrease oxygen in the media, it will result in mitigation of corrosion due to decrease in cathodic reaction rate. This was exhibited using bacterial biofilms of two bacteria *E. coli* and *P. fragi* on the basis of their ability to form biofilms and inhibit corrosion. Three media were chosen for this purpose aerobic/anaerobic namely seawater mimicking Vätäänen nine salts solution (VNSS), Luria-Bertani (LB) medium, and modified Baar's medium. The steel samples were tested for extent of corrosion experienced by them in these sterile/inoculated media in a test that continued for 2 weeks. It was observed that (i) in anaerobic solutions, corrosion rates remained unaffected irrespective of whether they were sterile or inoculated and (ii) in aerobic solutions, corrosion rates of steel were significantly decreased when exposed in inoculated media with respect to sterile. The reduction was more in case of inoculation of VNSS and LB with *E. Coli* than *P. fragi*. While in case of exposure in Baar's media, reduction in corrosion rate was observed more when solution was inoculated with *P. fragi* than *E. coli*. Thus, one observes reduction in corrosion rate only in aerobic media not in anaerobic media. Overall, the corrosion rate was found to decrease 4–43-fold in VNSS, 4–20-fold in LB and 20-fold in Barr's medium [16]. Further, the formation of biofilm over steel sample was verified by measuring open circuit potential (OCP) of steel in various aerobic/anaerobic media. OCP of steel, immersed in bacterial media, was found to decrease by nearly 200 mV after 8–10 days of cultivation. A comparable decrease in OCP, in the presence of biofilm-forming bacteria, has been reported by other workers as well. Formation of biofilm, over steel exposed in these test in bacterial medium, and their thickness was confirmed by confocal scanning laser microscopy (CFSL). Accordingly, *P. fragi* forms a more robust and thicker biofilm than *E. coli* which agrees well with a slightly larger drop in OCP for *P. fragi* relative to *E. coli*. The presence of biofilm reduces the diffusion of oxygen toward steel thus depletes the amount of oxygen in the vicinity of metal. Lesser amount of oxygen decreases the reduction reaction rate through a phenomenon called 'concentration polarization.' These changes lead to decrease in corrosion rate due to biofilm formation.

Bacteria Inhibiting Growth of Corrosion-Causing Bacteria by Producing Antimicrobials

Just now we discussed corrosion inhibition in the presence of aerobic environment using microbial route. The biofilm by the bacteria acts as a diffusion barrier for oxygen, which is responsible for main cathodic reaction, and so inhibit corrosion. However, there are anaerobic bacteria, e.g., SRB which imparts severe MIC on steel, Cu alloys, etc. In anaerobic environment, one may have to consider another mechanism, whereby MICI can be affected using microbial approach. One such approach is through some chemicals which are antimicrobials which inhibit the growth of bacteria, i.e., SRB responsible for causing MIC of metals. MIC on 1018 mild steel and SS-304 stainless steel was tested by exposing the sample in suspension inoculated with corrosion-causing anaerobic bacteria *SRB D. vulgaris* and *D. gigas*. The antimicrobials used were peptides *S. gramicidin* and *D. gramicidin,* and polymyxin B as well as the cationic peptides amidated and non-amidated indolicidine, bactenicine from bovine neutrophils [17]. These antimicrobials were observed to decrease the viability of both SRB's by 90% after a 1 h exposure at a concentration ranging from 25 to 100 ppm. Some peptide antimicrobials are relatively small, they are chosen for expression in biofilm-forming bacteria, e.g., SRB in present case. 'Gene expression' is a process by which the information contained within a gene becomes a useful product. So, to decrease corrosion by inhibiting SRB in biofilm, the genes for indolicidine and bactenicine were expressed in Bacillus subtilis BE1500 and B. subtilis WB 600 and the antimicrobials were secreted into the culture medium. Concentrated culture supernatants of *B. subtilis* BE1500 expressing bactenecin at 3 ppm decreased the viability of SRB *D. vulgaris* by 83% in suspension cultures. *B. subtilis* BE1500 and *B. subtilis* WB600 expressing bactenecin in biofilms also inhibited the SRB-induced corrosion of 304 stainless steel by 80–90%. This was accompanied with 97% decrease in the population of *D. vulgaris* in a *B. subtilis* BE1500 biofilm expressing bactenecin. The indolicidine construct was not as effective in inhibiting SRB on stainless steel as bactenecine construct due to the acid from the former.

Consequently, *B. subtilis* BE 1500 and WB600 were more resistant to bactenicin than to indolicidine.

Another attempt [18] was made for inhibiting mild steel corrosion from corrosion-causing anaerobic sulfate-reducing and aerobic iron-oxidizing bacteria by using gramicidin-S-producing biofilms. The bacteria chosen were *D. Orientis* ATCC 23598 (gram-positive SRB, formerly *Desulfovibrio orientis*) and *L. discophora* SP-6 ATCC 51168. In these experiments, the test media was a mixture of modified Baar's medium and ATCC 1503 medium consisting of the components in the amount as required for the growth of both the bacteria (*D. orientis* and L. discophora SP-6) when they were inoculated in the same corrosion flask. The medium was therefore named as DL medium. First experiments done were to see how bacterial counts of *D. orientis* and L. discophora, in the DL media, get affected by the presence of gramicidin (present in solution with several dilutions) indicating about of the survival of bacteria vis-à-vis concentration of gramicidin. Coupons were then exposed in the inoculated DL media having (i) pure gramicidin, (ii) P. Polymyxa 10401, which produces biofilm but not antibiotics, and (iii) a biofilm-forming B-brevis 18-3 which produces antimicrobes gramicidin-S. The corroded coupons were subjected to weight loss and biofilm thickness (using confocal microscopy) measurements. Also, electrochemical impedence measurement (EIS) was done on the above metal-environment systems.

Since, one has to consider effect on corrosion rate with respect to that observed in DL media inoculated with *D. orientis* and *L. discophora* SP-6 with and without *B. brevis* biofilm meant for introducing gramicidin in the media. Also, since one of the bacteria is aerobic it consumes oxygen and form a protective film, the decrease in corrosion rate may be confused with that due to this protective film. Accordingly, several variations in the composition of test media were considered. Let us compare the changes observed in corrosion rate under these circumstances group wise from case 1–6 (Table 4.4).

(1) **Case 1 and 2**: In sterile media, corrosion rate of mild steel is 93 μm/year, while in case 2, it is 36 μm/year. The decrease can be associated with the protective biofilm from *P. polymyxa* 10401, as it excludes oxygen. This results in decrease in reduction reaction rate which in turn reduces corrosion rate.

(2) **Case 2 and 3**: In case 2, due to protective biofilm of *P. polymyxa* 10401, the oxygen concentration near the metal will be low so corrosion rate is lesser. While in case 3, due to some oxide film because of exposure to air, oxygen will be lesser in amount near the metal resulting in anaerobic condition. This leads to growth of SRB which enhances corrosion. It is also supported by the observation of black surface on the metal surface of sulfide and smell of H2S near the test set-up. So corrosion rate in this case is higher than case 2.

(3) **Case 3 and 4**: Here, due to protective film formed as a result of PP, in case 4, the anaerobic environment produced near metal surface let SRB to grow in large amount. This leads to corrosion of mild steel due to SRB as observed from thick black deposit of sulfide and smell of H$_2$S same day after inoculation. In case 3, due to no protective film, concentration of oxygen is higher near metal surface and so degree of anaerobicity is less resulting in growth of SRB to lesser extent. Corrosion rate observed here is lesser when we compare EIS-based corrosion rate data. However,

Table 4.4 Effect of gramicidin and protective film (*P. polymyxa*) on MIC due to IOB and SRB

S. No.	Protective biofilm	Corrosion bacteria	Corrosion rate (μm/year)	EIS-based* corrosion rate
1	Sterile	–	93 ± 10	110 ± 46
2	*P. polymyxa* 10401	–	36 ± 4	66 ± 7
3	No protective biofilm	*D. Orientis*	71 ± 3	77 ± 11
4	*P. polymyxa* 10401	*D. Orientis*	60 ± 10	173 ± 54
5	*P. polymyxa* 10401	*D. Orientis* + *L. discophora*	111 ± 11	404 ± 137
6	*B. brevis* 18-3	*D. Orientis* + *L. discophora*	25 ± 10	20 ± 2

*This parameter represents corrosion rate—calculated from EIS data $R_p = \int(1/R_p A)dt$
R_p—Polarization resistance from EIS data as a function of time

the mass loss difference, which is the basis of calculating corrosion rate (μm/year), is not much.

(4) **Case 4 and 5**: In case 5, iron-oxidizing bacteria *L. discophora* has been added alongside SRB *D. orientis*. This IOB, itself or combined with biofilm-forming PP, has been observed not to cause much corrosion on mild steel in ATCC 1503 media. However, when IOB *L. discophora* is mixed with *D. Orientis* SRB, corrosion on mild steel increases significantly, in the test media, as compared to that in case of IOB alone (case 4—Table 4.4). This is evident from increase in corrosion rate by 1.8 times in case 5 as compared to case 4. This is also supported from the observation of enhanced growth of *D. orientis*, strong odor of H_2S, the black medium in the cell and thick black film on mild steel surface of iron sulfide.

(5) **Case 5 and 6**: This is the case where we observe the inhibiting effect of gramicidin-S on MIC due to SRB and IOB in the test media. Since pure gramicidin-S inhibited both *D. Orientis* and *L. discophora* bacteria, a B-brevis 18-3 biofilm was used to produce gramicidin-S from within the biofilm to inhibit both corrosion-causing bacteria simultaneously. One observes the corrosion rate decreases significantly to 25% (case 6) of the value observed without B-brevis biofilm (case 5). EIS-based corrosion rate value is still lower (~5% of the value without *B. brevis* biofilm). During the experiment, no H_2S odor was detected also the medium and the metal surface did not turn black. This indicates qualitatively that practically SRB did not grow in the presence of gramicidin-S. This protection, achieved in case 6, cannot be associated with the removal of oxygen, otherwise, higher corrosion rates may not have been observed in case 5 where media has both the bacteria and *P. Polymyxa* which forms protective biofilm obstructing the diffusion of oxygen toward metal surface. This clearly indicates the inhibiting effect, on MIC, of gramicidin-S produced by B-brevis 18-3 biofilm through thwarting the growth of both corrosion-causing bacteria.

There is another example of **mixture of nitrate-reducing, sulfide oxidizing bacteria** which inhibit SRB's thereby reducing corrosion due to SRB anaerobic media. SRB get energy required for their growth by a combination of redox systems involving metal oxidation and sulfate reduction. The reduction reaction produces sulfides responsible for SRB corrosion. The basic principle of inhibition involved in this alternative is by inhibiting SRB's population through a combination of sulfide/sulfate oxidation ($E^0 = -217$ mV) (2.1) and nitrate/nitrite-reducing ($E^0 = +$ 420 mV) reaction (4.4). Since standard redox potential for NO_3^-/NO_2^- reaction is cathodic and that of S^{2-}/SO_4^{2-} is anodic, sulfide will oxidize and nitrate will reduce. ΔG^0 for above oxidation-reduction combination reaction is –ive; hence, the oxidation of sulfide and reduction of nitrate will be a spontaneous reaction. In the presence of a metal, e.g., iron (as in instances of MIC of steel) whose $E^0 = -447$ mV, NRB will outcompete SRB because nitrate is strong oxidizer than sulfate for iron. This is so because potential difference between iron/nitrate couple is more ($\Delta E^0 = 867$ mV) than that between iron/sulfate couple ($\Delta E^0 = 664$ mV). Due to nitrate reduction reaction, redox potential of the system shifts to higher value which will inhibit SRB growth. Thus, the reduction in SRB and sulfides in the media will decrease corrosion significantly. Application of such a strategy is expected to be helpful in oil industry. Presence of sulfide makes crude oil 'sour' which enhances toxicity of crude and corrosion, e.g., pitting, sulfide cracking, etc., of pipe material. Reduction in sulfides and SRB through addition of nitrate-reducing bacteria is expected to address this problem effectively. Previously, some work [19] has been done on testing different nitrate-reducing, sulfate-oxidizing (NR-SOB) strains, e.g., Thiomicrospira sp. strain CVO, etc., to check extent of nitrite per nitrate reduction and of sulfide oxidation.

Microbial Inhibition of Corrosion Through Protective Layers

Another approach to inhibit corrosion by microbial means is to use microbes which produce protective layers, e.g., passive oxides formed and entrapped in biofilm matrix or the biomfilm matrix itself. Thus, it has been reported [20] that aerobic *Pseudomonas cichorii* is able to protect by inhibiting corrosion of mild steel in phosphate buffered basalt salt solution. The cause of protection was suggested as the formation of an iron oxide/iron phosphate formed within biofilm matrix. In this manner, several other microbial cases have been suggested which provide protection against corrosion in various cases. These and some other ways, e.g., anticorrosive influence of Acetobacter aceti biofilms on carbon steel, etc., of controlling corrosion/MIC through microbial route will be discussed in a later chapter.

References

1. Beech I, Flemming HC (2000) Microbiological fundamentals in "simple methods for the investigation of the role of biofilms in corrosion", Brite Euram Thematic Network on MIC Of Industrial Materials, pp 3–14

2. Sachan R (2020) Microbial influenced corrosion due to metal-oxidizing bacteria, Ph.D. Thesis, IIT Roorkee
3. ASTM Std. A. 2003. G1-03. Standard practice for preparing, cleaning and evaluating corrosion test Specimens. Annual book of ASTM standards, vol 3, pp 17–25
4. XRD Reference Data Card, Software xpert: for identification of solid compounds, Ultima IV Rikagu make X-Ray Diffractometer
5. Jack TR (2002) Biological corrosion failures. ASM Handbook, vol 11. Failure analysis and prevention, pp 881–891
6. Enning D, Garrelfs J (2014) Corrosion of iron by sulfate-reducing bacteria: new views of an old problem. Appl Environ Microbiol 80:1226–1236
7. Beech IB, Gaylarde CC (1999) Recent, advances in the study of bio-corrosion—an overview. Revista de Microbiologia 30:177–190
8. Lata S, Sharma C, Singh AK (2011) Microbial induced corrosion due to Desulfovibrio desulfuricans. Anti Corrosion Methods Mater 58:315–322
9. Lata S, Sharma C, Singh AK (2012) Microbial induced corrosion by thermophilic bacteria. Cent Eur J Eng 2:113–122
10. Lata S, Sharma C, Singh AK (2013) Effect of host media on microbial influenced corrosion due to *Desulfotomaculum nigrificans*. J Mat Eng Perform 22:1120–1128
11. Hamza E, Ibrahim Z, Abdolahi A (2013) Influence of *Pseudomonas aeruginosa* bacteria on corrosion resistance of 304 stainless steel. Corros Eng Sci Technol 48:116–120
12. Sachan R, Singh AK, Corrosion of steel due to iron oxidizing bacteria. Anti-Corros Methods Mater. https://doi.org/10.1108/ACMM-05-2018-1928
13. Boopathy R, Daniels L (1991) Effect of pH on anaerobic mild steel corrosion by methanogenic bacteria. Appl Environ Microbiol 57:2104–2108
14. Maruthamuthu S, Dhandapani P, Ponmariappan S, Bae J-H, Palaniswamy N, Rehman PKSM (2009) Impact of ammonia producing *Bacillus* sp. on corrosion of cupronickel alloy 90:10. Met Mater Int 15:409–419
15. Stadler R, Fuerbeth W, Herneit K, Grooters, Woellbrink M, Sand W (2008) First evaluation of the applicability of microbial extra cellular polymeric substances for corrosion protection of metal substrates. Electrochim Acta 54:91–99
16. Jayaraman A, Cheng ET, Earthman JC, Wood TK (1997) Axenic aerobic inhibit corrosion of SAE 1018 steel through oxygen depletion. Appl Microbiol Biotechnol 48:11–17
17. Jayaraman A, Mansfeld FB, Wood TK (1999) Inhibiting sulfate reducing bacteria in biofilms by expressing the antimicrobial peptides indolicidine and bactenicin. J Ind Microbiol Biotech 22:167–175
18. Zuo R, Wood TK (2004) Inhibiting mild steel corrosion from Sulfate-reducing and Iron oxidizing bacteria using gramicine-S-producing Biofilms. Appl Microbiol Biotechnol 65:747–753
19. Greene EA, Hubert C, Nemati M, Jennemen CE, Voodouw G (2003) Nitrite reductase activity of sulphate reducing bacteria prevents their inhibition by nitrate reducing, sulphide oxidizing Bacteria. Env Microbiol 5(7):607–617
20. Chongdar S, Gunashekharan, Kumar P (2005) Corrosion inhibition of mild steel by aerobic Biofilm. Electrochim Acta 50:4655–4665

Chapter 5
Industrial Cases of Microbial Induced Corrosion

Abstract Investigation on Microbial induced corrosion is of importance if it helps in overcoming problems posed by corrosion. It is, therefore, of utmost importance if such type of work can be used in controlling MIC experienced by the construction materials used in various industries or elsewhere. This chapter, therefore, deals with different industries where MIC is observed and pose problems. This chapter discusses about corrosion problems due to MIC in (i) oil and gas industry, (ii) pipeline in petro-chemical plant, (iii) nuclear power plant, (iv) high temperature environment, (v) pipeline and water systems, (vi) fire sprinkler systems, (vii) cooling cycles, (viii) sewage plants, (ix) marine environment and off-shore structure, (x) maritime vehicles, (xi) offshore infrastructure, and (xii) organic fuel.

Keywords MIC cases · Oil and gas · Nuclear power plant · Pipeline · Petrochemical plant · Sewage plant · Marine environment

A study or investigation in any area is worth pursuing if it has application for society directly or indirectly. Microbial induced corrosion (MIC) is a type of corrosion of metals when in contact with various biotic media primarily due to the microbes/bacteria present in them. Long construction time usually observed during building up of a plant, irrespective of any industry, and a large number of redundant or standby systems where water penetrates from surroundings, due to various factors including rains, and is allowed to remain stagnant for long periods of time produce conditions under which microbial induced corrosion can occur. Carbon and low alloy steels, stainless steels and copper alloys all are susceptible to MIC even in raw water application. As a result, MIC affects, more or less, every place having metallic/concrete structure. Consequently, MIC has its importance for different industries (Fig. 3.2) because the biotic media may be observed in many industries as process liquors whose function is to help in the process of conversion of raw material into a product, of course with the help of different chemicals at different stages of an industry, e.g., petrochemical complex which produce petroleum and other bi-products, oil exploration, and production facilities where oil is taken out from oil wells (offshore or on land) and is converted into various products from crude oil, power generation, drinking water, and sewer systems, maritime industry vehicles, e.g., boat, ships, submarines, and offshore structure etc. These process liquors interact with different type of materials (metals, plastic, concrete etc.) of different machinery parts of a factory or a plant as their material of construction. An industrial plant is likely to have different systems which may suffer from biocorrosion and/or biofouling, e.g., water injection lines, storage tanks, residual water treatment systems, filtration system, different type of pipes, reverse osmosis membranes etc. The nature of biofouling hence biocorrosion depends from one industry to another on the basis of the process which, in turn is governed by starting raw material and end product, type of bacteria which are likely to exist depending upon composition of its abiotic components, other environmental conditions etc.

Industrial Cases of MIC (including biodeterioration) can be outlined as below:

- Heat exchangers (failure, damage of ~55 million US$ stainless steel heat exchangers in 8 years. Under deposit pitting of heat exchanger tubing in nuclear power generating plants has been estimated to cost several hundreds of thousands dollars per unit per day in replacement energy cost.
- Water heater systems (MIC—life determining factor)
- Oil and gas—production, transportation, storage (100's of millions US$ per year in USA). In one oil company, 34% of corrosion damage was estimated to relate to microorganisms.
- Concrete sewers, biologically attacked textiles or decaying pieces of cultural property
- Ship building, harbor -installations, offshore platforms, bridges and buildings in places with marine environment, Maritime vehicles, e.g., ships, boats, submarines etc. (material exposed to natural environment)
- Power plants—here heat exchangers generally use river water/groundwater as cooling fluid
- Buried pipelines, oil refineries, and water treatment plants. MIC related cost of repair and replacing of piping material used in different systems has been estimated to be US$ 0.5–2 billion per annum.
- Petrochemical Plants.

As discussed in an earlier chapter that corrosion is of great relevance for a nation since it affects the economy of industrial production of that nation by prematurely and unscheduled restricting the useful life of industrial machinery. This may need maintenance and in ultimate case the replacement of machinery or its parts. This expenditure is represented as corrosion cost and a significant component of this is due to MIC. According to various factors outlined in Chapter 3, the cost of corrosion due to MIC can be estimated to be ~US$ 140 billion which is between 20–50% of the total corrosion cost. The actual cost at present must be much more than this figure. But easily one can say that MIC cost may be running in terms of 100's of billions of US$. It is, therefore, essential to check MIC occurring in different stages of industrial plants/factory, work on factors/circumstances responsible for its occurrence and take steps necessary to minimize the events of MIC to reduce the MIC cost. Keeping this in mind, discussion, industry wise, on various sections of industries affected by MIC, their causative factors, etc., have been detailed in the forthcoming part of this chapter.

5.1 Oil and Gas Industry

Microbial activity and so MIC are mostly observed, in this industry, in water handling system in oil fields and also in well bore, pipelines, and tanks. It imparts a huge economic impact on the oil producers due to the cost of leakage, unplanned shutdowns, maintenance, and chemical costs. Water handling systems consists of water injection system and produced water treatment facility presents conducive environment for the growth of microorganisms due to favorable temperature, organic and inorganic nutrients.

- Water injection systems on oil production platforms

The problem here is mainly due to two species of SRB's namely *Desulfovibrio sp.* and *Desulfobacter sp.* Seawater is made anaerobic before it is injected into the oil reservoir [1]. This environment is responsible for the growth of SRB's as they are anaerobes. These bacteria are found in anaerobic locations in the oil field, the reservoirs, wells, flow lines and other equipment. These bacteria are a great problem, most important being MIC of mild steel framework and souring of reservoir due to formation of H_2S. This gas is formed through a combination of redox reactions responsible for SRB MIC. These sulfides combine with Fe^{2+} ions to form different type of iron sulfides, from protective type mackinawite to more aggressive type greigite or pyrite, resulting in rapid corrosion such as pitting of mild steel pipes of the well.

While analyzing microbial corrosion in North Sea oil exploration, sulfate-reducing bacteria, SOB, hydrocarbon-oxidizing bacteria, iron-oxidizing bacteria, and slime forming bacteria and fungi were identified for causing corrosion [2]. The authors defined two distinct forms of SRB mediated corrosion:

(i) Pitting caused by SRB growing in the biofilm on metal surfaces,
(ii) Sulfide stress corrosion cracking (SSC) and hydrogen-induced cracking (HIC) or blistering caused by hydrogen permeation in high dissolved sulfide conditions.

Electrochemical tests have been performed on several stainless steels and mild steel in cultures of SRB. In all cases the results obtained with potentiodynamic polarization techniques suggested that pitting resistance was lower in cultures of SRB.

While examining effectiveness of a biocide to control MIC, one tests efficacy of biocides to minimize concentration of sessile bacteria although biocides are more effective on planktonic ones (In some cases, 5–10 times as much biocide was required to penetrate a biofilm than was required to kill planktonic bacteria). This is so because the sessile bacteria are responsible mainly for forming biofilms on metal surface hence MIC. It was, therefore, evaluated [1] to check the effectiveness of biocides to minimize the population of sessile SRB. Three different biocides required different contact time to achieve satisfactory penetration into biofilm and kill the sessile bacteria. Thus, biocide A required 24 h while biocide B and C required only 3–6 contact hours at the same concentration. Table 5.1 displays result of the test done

Table 5.1 Effectiveness of biocides to penetrate biofilm

S. No.	Biocide	SRB/steel coupon	Corrosion rate (mm/year)
1	A	10^6	>55
2	B	10^3	20
3	C	10^2	<5

with different biocides. The visual appearance of the steel coupons varying from severely pitted (A) and filmed to a totally uncorroded and non-biofouled (B and C) one for the least corrosive case.

Field evaluations carried out, in this connection, supported the outcome of laboratory results.

- SRB Induced Corrosion in an Oil Field

In an oil field, operating for several years, an increase of H_2S (sensed by rotten egg smell) was observed [3]. No serious corrosion effects were observed. However, formation of H_2S in water led to plugging of injection wells by FeS. Analysis of aqueous phase from different places of the system indicated that the crude oil processing unit (treater) was the main source of sulfides. The operating temperature of the treater, measured in oil phase, was 60–70 °C. The temperature in the water phase was 30–45 °C depending on the crude oil throughput and on the distance from inlet. To analyze, SRB's were taken from the water samples near the bottom of the treater. The samples were found to have FeS mainly and oil particles. Microscopic observation revealed oval-shaped particles mostly attached to the oil particles. Estimation of counting of different type of bacteria per mL of oil treater showed 6.3 × 10^6 *Desulfobacter* sp. as dominant species and 1.4 × 10^5 *Desulfovibrio* sp. alongside lesser amount of another bacteria similar to Desulfovibrio sapovorana and an unknown bacteria.

Now let us discuss the factors responsible for observing SRB's in oil well, in marine sediments etc. SRB in nature is expected whenever decomposable organic matter, e.g., cellulose is present alongside sulfate containing water where oxygen is limited. Typical habitats are aquatic sediments where settled organic particles accumulate. Significant activity of SRB is observed in salt marsh or marine sediments due to high sulfate concentration. Fermentative bacteria, e.g., *desulfobacter* sp., present in such places, cleave, and ferment the complex organic matter (cellulose, starch, and other biopolymers) to low-molecular-weight compound, e.g., acetate, H_2S, etc. Further, CO_2 is observed underground captured from atmospheric air. In natural anaerobic environments, the quantitatively most important fermentation products formed with CO_2 are H_2, acetate, propionate, and butyrate. Typical nutrients for SRB are simple compounds of low molecular weight, such as acetate, H_2S, propionate. Therefore, SRB in nature depend on fermentative bacteria. If sulfate is also present under anaerobic conditions, the fermentation products, e.g., acetate are used by SRB to get energy for their growth though redox reactions 5.1 and 5.2. These bacteria use sulfate as electron acceptor in their energy metabolism, a net oxidation of the organic substrates is effected without free oxygen, the reducing power from the decomposed organic matter appears as H_2S.

$$CH_3COO^- + 2H_2O \rightarrow 2CO_2 + 7H^+ + 8e^- \quad \text{Oxidation} \tag{5.1}$$

$$SO_4^{2-} + 10H^+ + 8e^- \rightarrow H_2S + 4H_2O \quad \text{Reduction} \tag{5.2}$$

$$\Delta G^0 = -41 \text{ KJ/mole of acetate}$$

Despite the inhibitory effect of oxygen on SRB, these bacteria are sometimes active in aerobic aquatic sediments, where SRB thrive in anaerobic microniches. Formation and maintenance of such microniches is explained by two factors. First, the respiration of aerobic bacteria scavenges oxygen and lead to production of sulfate (Eq. 5.3). This makes the solution more anaerobic which is favorable for growth of SRB.

$$H_2S + 2O_2 \rightarrow SO_4^{2-} + 2H^+ \quad \Delta G^0 = -41 \text{ kJ/mole of sulfide} \tag{5.3}$$

Second, SO_4^{2-} thus produced triggers further redox reactions 5.1 and 5.2. This lead to establishment of newer colonies of SRB which can protect themselves against oxygen. If organic matter increases, the anaerobic microniches can sometimes expand in a self-stimulating process.

Several studies [4] in past have established the relation between the SRB Desulfovibrio desulfuricans, grown in a lactate/sulfate medium, on the anaerobic corrosion of mild steel. Higher corrosion rates as well as the transpassive dissolution of Fe(0) or Fe(II) compounds to Fe(III) were observed in the presence of a bacterial culture. Another work relates to the pitting of stainless steel by the SRB which also found that the biogenic sulfides enhanced the passivity breakdown in the presence of chloride anions.

- MIC in Oil and Gas Pipelines

MIC has been observed to be responsible for most internal corrosion problems in oil transportation lines and storage tanks. Under these conditions, water gets stratified which prohibits corrosion inhibitors to access walls of pipe properly and oil and inhibitors undergo degradation. Bacteria play important role in degradation, while oil degradation consequently affect fuel system which in turn influences corrosion of pipeline.

The type of corrosion caused by microbes results in sharp-sided rounded pits and often pits within the pits. These are typically scattered randomly along the pipe whenever the line becomes water-wet due to (i) low velocity which results in water lying (ii) due to low spots in the line, and (iii) when water is more than about 30% of the liquid phase. The pitting occurs in the bottom of the line, and often occurs as linear strings of pits associated either with the oil–water contact phase or along the edge of sediment deposits. Water can stratify and collect in pipelines when the velocity is too slow or if the pipeline is operated in a 'stop-start' mode.

Corrosion causing bacteria commonly observed inside oil and gas pipelines are acid-producing bacteria (APB) and sulfate-reducing bacteria (SRB). SRB produce hydrogen sulfide, while APB generate acetic acid/sulfuric acid.

Oil Degradation Due to Microbes

Many microorganisms have the ability to utilize hydrocarbons as sole source of energy and these microorganisms are widely distributed in nature. This activity depends upon chemical nature of compounds in petroleum mixture and on the environmental details. Most important genera of hydrocarbon utilizer in aquatic environments are *Pseudomonas* sp., *Achromobacter* sp., *Micrococcus* sp., *Nocardia* sp.,*Candida* sp., *Vibrio* sp., *Acinetobacter* sp.,*Brevibacterium* sp.,*Rhodotorula* sp.,*Sporobolomyces* sp.,*Corynebacterium* sp., and*Flavobacterium* sp. Studies indicate that the concentration of available nitrogen and phosphorus in water are severe limiting factors for microbial hydrocarbon degradation. Hence, the addition of nitrogen and phosphorous containing fertilizers can be used to stimulate microbial degradation of hydrocarbon. Oxygen is absolutely required for hydrocarbon degradation, however, rate of anaerobic degradation at best is negligible in nature.

Storage tanks for oil are put underground. Underground conditions are usually limited in oxygen and it is of special concern if hydrocarbon degradation is possible under anaerobic conditions. The initial attack on hydrocarbon requires oxygen, but for the subsequent steps, anaerobic processes may degrade partially oxygenated intermediates further. Under oxygen limited conditions, accumulation of degradation products in the form of fatty acids occur. This has been observed in case of presence of SRB in the fuel. It is generally accepted that anaerobic degradation of hydrocarbons is a very slow process as compared to aerobic degradation. Over very long periods, however, it might be significant. Reports on the anaerobic degradation in natural ecosystems have suggested that nitrate or sulfate could serve as electron acceptors in place of oxygen. Extent of fuel hydrocarbon degradation at ambient temperature is less at high pressure than at atmospheric pressure. One can thus understand persistence of oil in deep-ocean environment for long time periods of time because it degrades very slowly.

Consequences of oil degradation in fuel system: Microbial activity leads to formation of solids which are combination of living or dead cellular materials, inorganic bi-products due to MIC. These materials may block fuel lines, injectors, pipes, filters, etc., and adding potential corrosion problems. By-products produced as a result of increased corrosivity due to MIC can often change the actual chemical properties of fuels in storage tanks and transporting pipelines. Gaseous by-products from microbial metabolism such as H_2S, SO_2 and CO_2 are often generated in such quantities that they can dissolve in the fuel, increasing the silver strip corrosivity index. The H_2S generation can also be a severe health hazard to the operating staff. Other metabolites, such as biosurfactants, can increase the emulsification of fuels.

Oil Degradation and Corrosion

The biodegradation of aliphatic than aromatic hydrocarbons, present in diesel oil, by *Pseudomonas* sp. occur within 3 weeks of incubation [5]. Thus, the rate of degradation of aliphatic hydrocarbon is higher than that of aromatic hydrocarbons. The rate of addition of oxygen (consumption of hydrogen) to various strains is given below:

Brucella sp. < *Sulfobacillus* sp. < *Thiobacillus* sp. < *Maraxella* sp. < *Gallionella* sp.

   Addition of oxygen will increase MIC due to oxygen respiration by aerobic bacteria, e.g., *Gallionella* sp.

   Enormous amount of money is being spent every year on prevention, monitoring, inspection, and land repair of the corrosion-related damages. Internal corrosion control programs usually involve chemical treatment with corrosion inhibitors. Details about this will be discussed in the next chapter.

5.2   Pipeline in Petrochemical Plant [6]

Pipelines and fittings are considered to be strategic components in petrochemical plants, especially when it comes to sustainable and safe operation. Among them elbows, mostly with angles of 90° or 45°, are the most used ones. Elbows can experience severe damage in the case of dramatic changes in the flow pattern. Carbon steels are widely used as materials of construction for the transmission of petroleum products and water. On the grounds that carbon steel is not remarkable as a corrosion-resistant alloy, external corrosion of buried carbon steel pipelines/fittings is always considered a major issue in petrochemical plants. The consequences of failures in buried pipeline/fittings are severe in most cases, owing to (i) repair and replacement of damaged pipes are costly and difficult since the damaged area needs digging into the ground. This implies more time, needed for maintenance operations, associated with longer downtimes for an installation, (ii) failure of pipes/fittings can cause environmental contaminations, imposing health risks to the public. Pipeline corrosion is influenced by soil resistivity and chemistry, temperature, pH, aeration, and microorganisms. For buried underground carbon steel pipelines several measures are taken to minimize the corrosion risk, e.g., applying coating and/or cathodic protection system. Any damage to them may result in catastrophic consequences.

   MIC of steels, in pipeline environment, is mainly associated with SRB together with iron-reducing bacteria (IRB), Fe/Mn-oxidizing bacteria, acid-producing bacteria. SRB-induced biocorrosion of the interior of pipes is related to low content of oxygen, together with stagnant liquid inside the pipelines, which are conducive for growth of SRB's. In addition, SRB can appear under deposits of soil, water, hydrocarbons, chemicals, etc.

   Now, we discuss a case of severe microbial induced failure of carbon steel elbows in a buried pipeline of amine treatment plant meant basically for removing CO_2 and H_2S from 'sour' natural gas. This massive failure was observed in less than 5 years of installation. Soil of the site contained 9% coarse (>76.2 mm) particles, an important factor from the point of view of its water retention, oxygen diffusion, and solute transport all of them significant for corrosion. pH of the soil was 8.5 which is significant to have diversity of microbes. The soil also contained Cl^-, which affects passive film formed on metal surface thus causing localized corrosion. Both the inner and outer surface were covered by red/yellow–brown and in some spots black rust layer. Elbow was heavily damaged with larger pits and perforations, on outer surface. Some of the larger pits had penetrated the sheet to make big hole. There was no pitting on the inner side.

SEM examination showed that outer surface, of the elbow, was covered with a porous biofilm consisting of complex communities/colonies of microorganisms and EPS. This can dissolve oxygen and create gradients of pH, resulting in localized forms of corrosion, such as crevice corrosion and pitting. The energy dispersive spectra (EDS) showed the presence of oxygen, sulfur, calcium, chloride, and iron. Presence of sulfur hints at the prevalence of SRB in the environment. Energy for the growth of SRB is obtained by sulfate reduction in the presence of natural organic compounds and steel surface which lead to corrosion of iron. XRD Analysis of corrosion products, taken from the surface of pipe, identified significant amount of iron sulfide, iron oxy-hydroxide (FeOOH) and iron oxide. Iron sulfide is a typical corrosion product which forms in case of MIC due to SRB and it is non-protective in nature. Consequently, the extent of corrosion in this case is much higher. Also, its color is black which is responsible for observance of some black colored part in the rust obtained from the pipe. Rest of the colors are due to iron oxide/oxy-hydroxide.

The inner surface of pipe, connected to elbow, was much less corroded and there was hardly any sign of localized corrosion. It was covered at some areas with a porous scale-like layer intermingled with some globular α-FeOOH particles. Other corrosion products were having plate-like micaceous crystalline features and honeycomb-like structures, both being typical morphologies of γ-FeOOH. Therefore, overall, the inner surface was covered with different iron oxide-hydroxides and possibly other oxide products. EDS analysis of this part of pipe did not show any sign of sulfur and chloride, unlike what was seen at the outer surface. It appears that the failure has started from the outer pipe surface, and corrosion products at the inner surface were caused by the liquid inside the pipeline.

Surface scratches from the corroded elbows and samples of soil from the corrosion area were inoculated in SRB-specific media and incubated in anaerobic conditions for 20 days. The black precipitation observed is created due to SRB growth and H_2S and FeS production. It is noticeable that the SRB-induced black precipitation was only created on the corroded surface of the failed elbow, and there was no sign of black precipitation on the cut surface. The formation of SRB cells was confirmed by SEM.

It can, therefore, be concluded that severe corrosion (large and deep pits with one turning in a hole through pipe from outer surface) observed in the present case can be attributed to the presence of SRB's on the outer soil adjacent to the pipe. Corrosion observed on the inner part of pipe was much less due to the chemistry of the inner environment and without any significant role of microbes.

5.3 Nuclear Power Plant [7]

Nuclear power plants are particularly susceptible to MIC due to their basic design philosophy. MIC is observed under stagnant conditions or operation with low or intermittent flow. A large number of standby and redundant systems in nuclear plant design establishes those conditions in a number of systems, some of which are safety related. The large size of nuclear power facilities and the often-prolonged licensing process extend nuclear plant construction schedules, which leads to long stagnant periods, often with structural materials in contact with untreated water used for hydrostatic testing. When it starts operating, it utilizes water sources rich in organics, such as man-made lakes or cooling ponds, which may contribute to MIC initiation.

MIC in nuclear power industry is different from others in the sense that here SRB are not a major culprit for MIC. In some cases, intermittent flow appears more damaging than stagnant or low flow conditions, because sulfide rich, non-protective films formed during the stagnant period can be washed off by flowing oxygenated water. Corrosion, like the case of erosion corrosion, proceeds rapidly until a new film forms. Water from cooling ponds, artificial lakes, rivers, seawater all are potential source of providing conditions for bacterial growth. And all the structural materials, e.g., carbon steel, low alloy steel, stainless steel, copper-nickel alloys, etc. are susceptible to MIC. In addition to above mentioned conditions, carbon steels in power generation are amenable to corrosion due to hydraulic effects. This is caused due to thick deposits of corrosion products formed on pipe interiors which reduces the available flow area significantly even to the extent of blockage of pipes. Consequently, (i) flow of cooling liquid to critical equipment is insufficient and (ii) pressure drop increase. Removal of the deposit (mainly done mechanically) alleviates the flow difficulties. Unless the cause of the corrosion problem is removed, deposits will begin to form rapidly due to accelerated attack of the pipe wall jeopardizing the integrity of the pipe. Generally, corrosion deposits (or tubercles) comprise corrosion products from upstream locations, microbes, their mucilaginous exopolymer and other debris. In some situations, however, the bulk of the iron in the tubercle can come from the corrosion occurring beneath the tubercle itself (case of under-deposit corrosion). In this latter case, removal of the tubercle without destruction of the iron-oxidizing organisms can increase the corrosion rate and cause damage as fresh metal is exposed under aerobic conditions. In many environments, where MIC is a problem, corrosion in sterile water of the same chemistry would also be high. In these environments, the presence of microbes exacerbates the corrosive effects by producing much more rapid local perforations of the pressure boundary and the potential for flow blockage.

18–8 Stainless steels are commonly used in nuclear power plant construction in applications requiring corrosion resistance and high purity aqueous environment, e.g., reactor coolant system, emergency systems, reactor auxiliary systems, and, in many plants, in the feedwater train and condenser. They provide the added benefit of high toughness and demonstrated leak-before break margin in wrought and welded forms. Stainless steels may be subjected to microbial-influenced attack in high purity environments with microbial activity. The MIC of stainless steel is characterized by pitting, most commonly at weldments. Through wall pitting with resultant small leaks is the most common consequence of MIC of stainless steels. Iron-oxidizing bacteria, e.g., *Gallionella* and *Siderocapsa* are responsible for MIC attack on stainless steel. In case of seawater-cooled power plants, the biofilm formed over stainless steels, during seawater exposure, causes its corrosion potential to increase significantly. If the corrosion potential exceeds its pitting potential, the materials can be expected to pit. Since welded regions are often anodic to the base metal, pitting is anticipated at the welds. In such cases, the place of attack may not be necessarily at the microbial activity site. The 6% molybdenum stainless steels (SS254SMO, 654SMO) exhibit superior resistance to pitting in highly oxidizing chloride solutions. Laboratory demonstrations and service experience in raw waters also appear to bear out a resistance to MIC.

Copper alloys such as 70–30 and 90–10 copper-nickels, brasses, aluminum bronzes, and admiralty brass are used in nuclear plant service in pumps and valves and in various heat exchangers, ranging from small coolers to feed-water heaters to main condenser units. All of them are susceptible to MIC although they exhibit superior resistance to fouling.

Titanium appears to be the most MIC-resistant metal examined to date. Its extremely stable oxide film is highly protective in a variety of environments including all local environments established by microorganisms. However, titanium is susceptible to biofouling and its use may thus be limited. The expense associated with specification of titanium is great, although not prohibitive as evidenced by the increasing use of the material in critical heat exchanger applications, including the main condenser.

5.4   High Temperature

So far this chapter has dealt with instances of MIC in various industries/environment at room temperature. Considering the important role, bacteria and other microorganisms play in affecting the useful life of industrial machinery through MIC at room temperature, it is imperative to check their role at higher temperature too since many bacteria are known to exist at such temperatures. Accordingly, a case of MIC at higher temperature is being described here. This is the instance of MIC in geothermal electric power plants where its condenser works at a steam temperature ranging from 40 to 150 °C with chlorides, sulfates, iron, etc. Geothermal energy is derived from inside the earth where temperature could be very high so that underground water comes in the form of steam. Sulfur-rich steam coming out of the well is cause of many corrosion problems. Stainless steel pipes coming out of the well are observed to have pits. The observation suggests MIC could be contributing to the degradation in the condenser. Tubes and supports inside the condenser are of 304L SS. Bacteria identified in the condensate were *Desulfotomaculum nigrificans* and *Desulfotomaculum acetioxidans*. The tubes exposed to condenser environment for 4–8 months showed pitting.

5.5 Pipeline and Water Systems [8]

A pipeline is basically designed to transport liquid, gaseous or solid materials where water/aquatic media can often stagnate or move with low flow rate. This condition becomes a potential breeding ground for the development of microorganisms. These microbes attack metallic pipes by corrosion named as MIC. This biofilm attack of pipeline and water systems are of concern for various industrial processes. Biofouling occurs worldwide in various industries, from offshore oil and gas industries to cooling systems. Biofouling is also observed in heat exchangers, water-cooling pipes of power stations or factories. Just like a clogged drain in our kitchen or bathroom affects the flow and one needs to clean it, buildup of matter inside cooling system pipes decreases performance. The impact of such attacks range from mechanical blockages, reduced flow capacity, pipe wall penetrations and leaks. Table 5.2 gives a bird's eye-view of areas affected by biocorrosion in such systems.

The food industry is very much concerned with drinking water system due to contamination in pipes and process equipment caused by microorganisms attached to solid surfaces. Biofilms represent significant health risk because they can harbor pathogen and direct contact can lead to food contamination. Biofilms are notoriously resistant to many disinfectants, which often react with extracellular polymeric substances (EPS) and other organic constituents that form the biofilm, thus inactivating the disinfectant. Disinfectants may not reach bacteria in the biofilm that may proliferate even after disinfection. Thus, the effectiveness of a disinfectant used for biofilm control can be determined by measuring the rate and depth of disinfectant penetration. Most investigations indicate that microbial corrosion does not occur in absence of biofilms as evidenced by the presence of slime and tubercles caused from biofilm formations. Observations and analyzes of inner pipe walls at

Table 5.2 Systems with persistent MIC problems

Application/system	Problem component/areas	Microorganisms
Pipelines/storage tanks (water, wastewater, gas, oil)	Stagnant areas in the interior exterior of buried pipelines and tanks, especially in wet clay environments Inadequate drying after hydrotesting	Aerobic and anaerobic acid producers Sulfate-reducing bacteria Iron/manganese oxidizing bacteria Sulfate oxidizing bacteria
Power generation plants	Heat exchangers condensers	Aerobic and anaerobic bacteria Sulfate-reducing bacteria Metal-oxidizing bacteria
Cooling systems	Cooling towers heat exchangers Storage tanks	Aerobic and anaerobic bacteria metal oxidizing bacteria Slime forming bacteria Algae Fungi

biological corrosion sites show attachments of tubercles and their distributions of organisms to the metallic surfaces that cause corrosion, blockage and pitting holes in the pipes leading to failures. Microbial communities in local depositions through various mechanisms associated with differential oxygen cells, under-deposit chloride concentrations, sulfide attacks, and aspects of acid production.

Pipelines with pH of water ranging from 4 to 8 and temperature 20–45 °C are ideal for growth of microorganisms. Anaerobic bacteria, e.g., *Desulfovibrio*, *Desulfotomaculum* and *Desulfomonas* and *aerobic bacteria*, e.g., *Acidthiobacillus thioxidans*, *Acidthiobacillus ferroxidans*, *Gallionella*, *Siderocapsa*, *Leptothrix*, *Sphaerotilus* and *Pseudomonas* are observed in this media. Additionally, bacterial development also depends upon surface energy, surface roughness of metal, and fluid flow rate. Thus biofilm development is amenable to maximum flow velocity of 1.5 m/s. Beyond this the fluid force is sufficiently high to avoid sticking of bacteria on surface.

In one of the investigations on the impact of stagnant water on MIC, a water pipeline of galvanized carbon steel was tested. For the purpose of galvanic effect, this pipe had a zinc layer (hot dip galvanizing) ranging from 35 to 90 μm. Ordinary drinking water was allowed to flow for 2 years through the pipe. In between eight heat shocks with water temperature higher than 70 °C and several chlorine shocks were carried out in water supply systems. Inspection of pipe interior, after the test, showed coating of biofilm. Corrosion was more intense at the lower part of the pipe indicating that corrosion was due to stagnant part of water. The stagnant water was very appropriate medium for the microorganism growth and further development of MIC. They dissolve zinc and iron and even 304 stainless steel. Color of deposits in corroded part was reddish brown (MIC product due to IOB's) and black (due to SRB-induced corrosion). On removal of biofilm, one could smell of rotten eggs indicating H_2S gas produced by SRB's. EDX analysis of corrosion products showed presence of Fe, Zn, S, and O. The black colored product observed in deposits could be FeS and ZnS which are characteristics of MIC due to SRB's. The upper part of the exposed pipe showed white colored corrosion products which were identified, by analysis of EDS results, due to mainly Zn and O and to some extent due to Fe and O as no sulfur was found to be present as per EDS analysis. It is well known that corrosion of galvanized steel involves mainly oxidation of Zn to save iron oxidation so that overall corrosion resistance of galvanized steel is better than ordinary carbon steel. The analysis of water samples—one from the city water supply system and second from the pipe outlet was performed. The former sample did not show Fe and Zn while the latter showed significant amount of these chemicals. This basically indicates that the presence of Fe and Zn was not due to contamination of water from the city supply system but was due to SRB and IOB-influenced corrosion of pipeline material. One can conclude two things from these findings (i) MIC could lead to contamination of water which may be unfit for consumption and (ii) MIC leads to weakening of water supply pipe shortening its life which may require premature replacement of pipeline hence additional cost on the municipality of that particular area. The work also suggests two important points while designing such a system (i) the pipelines should be horizontally fitted such that they are self-draining type and

(ii) they should be designed in such a way that velocity of water/fluid flow is at-least 1.5 m/s.

5.6   Fire Sprinkler System [9]

Signs of MIC affecting fire sprinkler systems starts appearing from the beginning of 1990's because of observation of rapid development of (i) pin hole sized leaks and (ii) highly obstructive bacterial growths in the interior of pipes. Most of these occurred well before the life expectancy of the system, after serving for 5–20 years. In extreme cases, the critical obstruction and the pin hole leaks start appearing in less than one year. Since then, MIC in fire protection system has become an important part of MIC studies and discussion. As a consequence of MIC attack, the inner surface of pipe in a sprinkler system becomes rough with pits and valleys. This roughness introduces friction for the flow of water in pipe. So, pipe surface texture is critical in sprinkler system. An increase in surface roughness increases the pressure loss per ft of sprinkler pipe length. Even a small amount of internal corrosion may make a system ineffective in fire control. The smaller the pipe diameter, the more dramatically corrosion affects flow of water since then friction resistance is quite high. To avoid this effect, slightly larger diameter of sprinkler pipe is provided. This is like 'corrosion allowance' as we understand in general corrosion terminology. Ultimately, the fire protection system is likely to fail prematurely due to MIC. In this industry, a normal life expectancy for a sprinkler system, with proper maintenance, is expected to be 50 years before major repairs become imminent.

Following types of premature failure are expected in these systems:

(i)   The system is not able to hold water which could be due to presence of pin holes in the system. The system will fail functionally as it will not be able to hold water sufficiently to extinguish fire. In one case, MIC-related system failure resulted in shutdown of an aerospace manufacturer's global computing center. Water from a pinhole leak, developed in wet pipe sprinkler branch line located over the mainframe equipment of the computing system, damaged the computer and also resulted in loss of 5000 h of operational time.
(ii) The worrisome of the two—unable to achieve fire control. The failure not only affects property loss but also may threaten life of personnel working there.

There are only few studies done on the MIC related problems in fire protection industry. However, there is some information available on this aspect. Thus, in some countries, this industry accepts MIC as a wide spread problem in their system. Corrosion is 5th leading cause of system failures. In another study done in the late 1990's, about 150 cases of sprinkler system leaks were observed. MIC was found to be present in 40% of these cases. In later studies, on piping field samples, over 60% of failures were attributed to MIC.

Mostly, steel sprinkler pipes are observed to fail first. Sprinkler orifice caps, control valves, fittings and supply tanks have been found to have got damaged from MIC.

Many cases of obstructive growth and pinhole leaks, associated with MIC, have been found within several feet of the discharge site of fire pumps due to velocity of water flow. Steel and stainless steel are susceptible to MIC but with varying degree of resistance to MIC. Plastic components, e.g., underground water mains although do not get corroded but they are affected by biofouling or bacterial debris blockage from upstream corrosion activity. Source of bacteria in the system is not just its water supply but others too like soil, air, and cutting oils. It is not necessary that a system should be water filled to observe MIC, even dried system also observe MIC due to moisture present in the system after it has been drained.

5.7   Cooling Cycles [10]

A cooling system is one which keeps some area of an industrial part cool and dry either by way of passing cold air or by using a fluid to transfer heat from one place to another. The open circulated cooling system contains organic and mineral nutrients which provide excellent environment for growth of bacteria, algae, and fungus. Uncontrolled, they may lead to biofouling and unleash corrosion. Together with scaling, they affect the efficiency of cooling system.

Corrosion of mild steel, stainless steel, and brass has been evaluated using corrosion coupons placed in a coupon rack connected to the circulated water line. Continuous on-line corrosion rate and pitting tendency index measurements is done by linear polarization resistance technique.

The type of nutrients that are present in the recirculated cooling water govern the type of bacteria likely to be found in the cooling circuit. Nitrite and nitrate-oxidizing bacteria and SRB's are mainly found in the cooling system. In the ammonia and urea plants, a common contaminant of the cooling water is ammonia. This contaminant may enter in the cooling circuit either from process fluids—by leakage or from atmosphere—by absorption. Even small quantities of ammonia (~ppm) are sufficient for inducing the growth of bacterial mass. In the nitrification process, which is strictly aerobic, ammonia at pH < 11 changes to ammonium ion (Eq. 4.11). Ammonium ion (NH_4^+) oxidizes to nitrite in the presence of ammonia oxidation bacteria (nitrite forming bacteria) *Nitrosomonas* genus. Nitrite oxidizes to nitrate ion by nitrate forming bacteria *nitrobacter* genus (Eq. 4.4). e^- thus produced are consumed by oxygen reduction (Eq. 1.5a) thereby maintaining electrical neutrality. Energy derived from above redox reactions help in growth of both bacteria. The optimal conditions for nitrification are: 15–400 ppm NH_3, pH = 7.5–8.0, $T = 298$–313 K. The carbon source for bacteria is represented by the carbon dioxide and carbonates dissolved in the cooling water. It is observed that the increase of nitrate concentration, that denotes an increase of the nitrifying bacteria population, lead to higher corrosion rate and increased pitting.

Another type of bacteria observed are Sulfate-reducing bacteria (SRB) *Desulfovibrio, Desulfobacter*, etc., genus which are anaerobic. As per the reactions described in earlier chapter, SRB's reduce the sulfates into H_2S. Source of sulphate ions (850–1200 ppm SO_4^{2-}) in the system is H_2SO_4 which is used to control pH. Due to the use of H_2SO_4 as pH regulator, the recirculated water has an increased content in sulfate ions than in those systems that use no pH regulation (295–310 ppm SO_4^{2-}). SRB colonies have been reported in the mud from the cooling tower basin. Carbon steel coupons are observed to show pits of hemispherical shape, which is characteristic of pits due to SRB's.

In some cases, one also observes slime forming bacteria. These bacteria are observed in systems where contaminants from the process sides are hydrocarbons because of cooling system being part of fluid cracking unit. The slime forming bacteria secrets variety of extracellular polysaccharides (EPS) which form biofilm, a slimy layer. Slimy layer reduces the diffusion of oxygen toward the bottom of the layer, which may be in contact with metal surface of a process machinery. Area beneath the biofilm thus transits from aerobic to anaerobic environment leading to growth of SRB's and acid-producing bacteria. In addition to MIC, the slimy layer is also responsible for reduced thermal conduction, hydraulic pressure leading to clogging of pipes.

5.8 Sewage Plants [11, 12]

Sewage treatment is the process to remove contaminants, which otherwise may be harmful for human consumption, agriculture, water life, e.g., fishes etc., or in the case of industrial operations removal of chemicals which otherwise may be harmful for other industrial processes. The sewage treatment plant is, therefore, meant for treating municipal wastewater containing mainly household sewage, industrial waste water meant to remove the contaminants of an industry. The treatment process uses physical, chemical and biological methods to remove the contaminants and produce treated wastewater that is safe enough either for release into the environment, e.g., around water bodies, river, plantation etc., or for recycling into the industrial system. The sewage treatment plant consists of stages which together remove all types of contaminants. These are (i) **filtration** by bar screen for removing large sized impurities to remove them physically (ii) **aeration** for growth of bacteria which are present in sewage. The growth leads to flocculation which then may float to top known as 'creaming, or settle at the bottom of the tank as 'sediments' or be readily filtered. The process helps in removing bacteria. (iii) **PAC dosing** is addition of a given amount of poly aluminum chloride which is meant for coagulation of pollutants that helps in their removal from sewage. (iv) **chlorine dosing** for disinfecting to remove chemical and biological hazards from sewage/waste water. 'Pressurized sand filters' are used for removing turbidity/suspended particles and 'carbon filter' for removing contaminants and impurity by chemical adsorption. All these stages are responsible for bacterial corrosion on metal and concretes structures/equipment of sewage treatment plant.

Laboratory test was conducted to investigate the MIC on carbon-steel in presence of IOB as these are important from standpoint of MIC in a typical sewage plant. For isolating the bacteria responsible for MIC of steel, a sterile metal loop is suspended inside the pipeline of sewage treatment plant for several days. Corrosion products formed on the metal (carbon steel) loop are collected to isolate the bacteria. Morphological and biochemical characteristics of the bacteria are studied as per standard procedure to identify the strain of bacteria. Thus, two novel iron-oxidizing bacteria (IOB) were identified. For corrosion evaluation, carbon steel was submerged in the IOB inoculated test solution, Bushnell Hass medium [$MgSO_4.7H_2O$ 0.2 g/l, $CaCl_2$ 0.02 g/l, NH_4NO_3 1 g/l, K_2HPO_4 1 g/l, KH_2PO_4 1 g/l, $FeCl_3$ 0.05 g/l] at pH = 7. After exposing for 1 month under freely corroding conditions, the carbon steel samples were examined for their surface characteristics and corrosion features by SEM after sputter coating with gold.

Electrochemical polarization tests show decrease in corrosion potential and increase in corrosion current density on adding bacteria to the sterile test solution. Results also show increased cathodic slope, after the sample is exposed to test solution longer, which has been attributed to decrease in concentration of oxygen. As is known, IOB attacks metal by oxidizing Fe^{2+} to Fe^{3+}, the electron thus obtained accelerate oxygen reduction to convert it into OH^- ion, following reactions vide Eqs. 1.1, 1.5a, and 1.7 due to IOB. Because of additional oxidation reaction 1.7, due to IOB, more oxygen gets reduced to neutralize additional e^-. When metal is exposed in the IOB inoculated test solution longer, more of Fe^{2+} will be oxidized to Fe^{3+} and hence more of O_2 will reduce due to availability of higher number of electrons. This decreases the concentration of dissolved oxygen in test solution, thereby shifting E_{corr} to negative values. Further, decrease in amount of oxygen also decreases rate of cathodic reaction which in turn is observed as increase of cathodic slope of the curve indicating decrease of rate of cathodic process. Electrochemical impedence spectroscopy (EIS) measurements indicate higher impedence in case of sterile test solution due to the thicker film formation of corrosion products. However, for IOB inoculated test solution, decrease in impedence is observed. This decrease could be due to ease of diffusion of electroactive species through the biofilm and oxide formed on the metal surface in the presence of bacteria. SEM images of the corroded samples show their surface covered with dense deposit but their pattern is quite different. While deposit on sample exposed to sterile media shows dense deposits, those exposed to media with IOB show brittle deposits. SEM result show grain boundary and localized attack on samples exposed to media in the presence of IOB. This could be attributed to formation of differential aeration, due to respiration by bacteria and presence of biofilm, and concentration cell in the presence of bacteria. XRD of the corrosion products formed on the corroded samples show only FeOOH in sterile as well as inoculated media. Probably the nature of FeOOH formed may be different due to observance of difference in the intensity of diffraction peaks in XRD pattern. It is necessary to know whether FeOOH formed in these cases was α-, β-, γ-, or δ-type.

Another important material of construction in sewage plants is **concrete** and these are also affected by MIC. Treatment plants are most of the time partially filled with sewage. Consequently, the space left above the water line, in the plant, is available for bacterial growth and accumulation of gas from sewage decomposition. Due to unusual bacterial proliferation, enormous bacterial concentration develops in short time representing a huge potential threat to stability of the structure due to MIC. The bacteria observed in plant is of genus *Thiobacillus* which grow and develop best at 25–35 °C. They are able to oxidize elemental sulphur, sulfide, thiosulphate and polythionates (Eqns. 5.4–5.7).

$$S + 4H_2O \rightarrow H_2SO_4 + 6H^+ + 6e^- \tag{5.4}$$

$$S^{2-} + 6H_2O \rightarrow SO_3^{2-} + 3H_2O + 6H^+ + 6e^- \tag{5.5}$$

$$S_2O_3^{2-} + 5H_2O \rightarrow 2SO_4^{2-} + 10H^+ + 8e^- \tag{5.6}$$

$$S_2O_6^{2-} + 2H_2O + O_2 \rightarrow 2HSO_5^- + 2H^+ + 2e^- \tag{5.7}$$

Bacteria of the Thiobacillus species can be divided into two groups (i) those that only grow at neutral pH values. They are responsible for the conversion of the elemental sulfur to sulfuric acid (Eq. 5.4) which is aggressive to concrete, steel, cast iron etc., and (ii) Others, e.g., *Thiobacillus thiooxidans* grow at acidic pH values (pH 2–5) under aerobic conditions. *Thiobacillus intermedius* is most active at pH 3–5 and oxidizes thiosulfate ions ($S_2O_3{}^{2-}$). Another one of this group *Thiobacillus ferrooxidans* develop in aerobic conditions and has a range of pH growth from 1.5 to 5. It oxidizes Fe^{2+} to Fe^{3+}, the e^-'s thus produced increase reduction reaction rate resulting in increased corrosion. As one can see that all reactions 5.4–5.7 increases acidity of solutions. Consequently, the local environment becomes so acidic that it causes dissolution of concrete and corrosion of steel and cast iron. In this process insoluble iron oxy-hydroxide forms which clogs steel/cast iron pipes.

Since sludge layers reduce oxygen passage significantly, deposition of sludge over concrete/metal pipes or other structures generates anaerobic environment in the sewage pipelines surface and on other surfaces of sewage systems. Hence, one observes presence of SRB's in these areas. SRB reduces sulfate ions to sulfide ions which react with hydrogen ions in the sewage to produce hydrosulfide, also called bisulfite ion (HS^-) which transforms to H_2S at temperatures above 15 °C. Overall, this leads to enhanced corrosion.

Presence of sulfuric acid (due to sulfur bacteria) results in drop of pH from 13 to even 6 (reactions 5.4–5.7) in sewage system which destructs the concrete tank by neutralizing reaction. Additionally, nitrification/denitrification concrete tanks are destructed by nitrifying bacteria, namely *Nitrosomonas, Nitrobacter, Nitrospira and the Nitrosococcus* species, due to formation of nitric acid by first oxidizing ammonium ion to $NO_2{}^-$ (Eq. 4.2) which in turn oxidizes to $NO_3{}^-$ (Eq. 4.4). Both reactions make solution acidic and electrons produced in the reactions are utilized by oxygen reduction reaction (Eq. 1.5a). The biogenic acid released in the process of nitrification acts more aggressively on concrete. Soluble salts of calcium are formed as a consequence of neutralization reactions leading to destruction of concrete tanks. Additionally, steel reinforcement is vulnerable to the action of the aggressive environment. Prolonged action of a weak acid will be indicated by occurrence of rust (corrosion of the reinforcing steel), cracking and spalling of the concrete. With aggressive acid action, there are no cracks in the concrete or spalling and only loose aggregates remain there. In many cases, concrete structures are damaged by microbiological corrosion to the point at which necessary repairs should take place after four years, which was followed by a total destruction after six years. In order to protect concrete against microbiological attacks, it is necessary to apply coatings that will make a barrier and will be resistant to the effect of biogenic inorganic acids (sulfuric, nitrate) and organic ones.

5.9   Marine Environment and Offshore Structure

The ocean covers about 70% of the earth surface area and ocean transport supports 90% of the freight transportation globally. It is also due to the fact that ocean freight transport is economical as compared to the air or road transportation. This, in turn, has made marine resources and marine industry one of the indispensable pillars in economic development. However, marine environment is extremely harsh corrosive environment due to presence of humidity, chloride and other salts (Table below), and marine organisms which together make seawater an electrolyte with high conductivity. Thus, seawater as an electrolyte affects severely the metals, etc., material of construction for maritime vehicles, industrial equipment/machinery, and infrastructure, through MIC and corrosion. MIC has been a persistent problem for maritime vehicles, equipment for deep-ocean exploitation, and under water exploration. Marine energy, deep-sea survey, port, and other infrastructure constructions need a large number of marine platforms, offshore wind power, vessels, pipelines, deep-sea storage equipment, and shore facilities, the construction of which requires various types of steel being able to operate in anaerobic or anoxic conditions.

Seawater has about 96.5% water and 2.5% salts and smaller amount of other inorganic and organic substances. Some of the important constituents of seawater (as g/kg of seawater) are given below.

Constituents	Cl^-	Na^+	SO_4^{2-}	Mg^{2+}	Ca^{2+}	K^+	C(Inorganic)
Amount (g/kg of seawater)	19.16	10.68	2.68	1.28	0.41	0.39	0.03

Additionally, seawater has carbohydrates and amino acids, P and N which act as nutrient for microorganisms' growth. Microorganisms involved in MIC processes in the marine environment include fungi, archaea, algae, and various bacteria. The miscellaneous groups of marine bacteria involved in iron biocorrosion can be divided into sulfate-reducing bacteria (SRB), sulfur-oxidizing bacteria (SOB), iron-reducing bacteria (IRB), iron-oxidizing bacteria (IOB), acid-producing bacteria (APB), slime-producing bacteria (SPB), etc., among which SRB is considered to be the main culprit of MIC due to its efficient electron transfer rate. Especially, the loss caused by SRB-induced MIC of steel materials accounts for more than 50% of the total MIC loss.

SRB play important role in facultative/obligate marine environment, e.g., in seawater, oil fields, etc., Long-term seawater immersion tests show *Desulfovibrio* sp., *Desulfobacter* sp.*and Desulfotomaculum* sp. as dominant SRB's in the inner surface of rust layer formed on steel (steel biofilm interface). At this interface, carbon sources (coming from seawater) of oxidation and oxygen content are insufficient. In the marine environment, microbes and their metabolites can significantly change the microenvironment of biofilm-steel interface to accelerate MIC process. One of these is that the polysaccharide in biofilm traps nutrients from outside which are meant for nourishing the bacteria inside. Consequently, the bacteria will look for alternative source of energy for its growth. The biofilm-steel interface microenvironment changes from aerobic to anaerobic which may lead to the formation of interactive metabolic homeostasis conducive to corrosion within the biofilm communities.

The EPS within biofilms serve as a glue to bind corrosive bacteria on steel surface and facilitate metabolism/respiration optimization. Modern techniques in molecular biology help in tracking in situ changes of structure and composition of microbial communities with time and space. Thus, previous immersion studies have shown that *Proteobacteria* (long-term tests) and *Bacteroidetes* (short-term tests i.e. <40 days) are the dominant bacterial phyla of biofilm on corrosion samples from marine facilities. *Proteobacteria* are the largest bacterial phylum found most abundantly in marine environment and many members of this phylum are pioneer colonizers and versatile biofilm builders. Most *Proteobacteria*, e.g., SRB in Deltaproteobacteria, etc., are associated with redox cycles of iron and sulfur [13].

According to dynamic corrosion model of steel in marine environment, corrosion is controlled by the reduction of dissolved oxygen (DO) in the early stage, leading to deficiency of oxygen with time, after which anoxic stage follows. It has been experimentally verified that carbon steel gets aggressively corroded by SRB after DO is exhausted. As explained by biocathode sulfate reduction (BCSR) theory, this happens via oxidation (hence corrosion) of iron (Eq. 1.1, $E^0 = -447$ mV) and respiration of electroactive bacteria SRB by reducing SO_4^{2-} to HS^- (Eq. 5.8) through the direct electron transfer (DET). This process is termed as "electrical microbial influenced corrosion" (EMIC) or Type 1 MIC. As discussed in an earlier chapter, the need of transportation of electrons arises because electrons are generated by oxidation of iron whereas sulfate reduction occurs inside the cell cytochrome. So electrons have to be transported from iron substrate to inside the cell [14].

$$SO_4^{2-} + 9H^+ + 8e^- \rightarrow HS^- + 4H_2O \quad E^0 = -217\,mV \quad (5.8)$$

This transfer is facilitated by so-called 'electron mediators' which are soluble compounds and are redox active (undergo redox i.e. oxidation and reduction).

Table 5.3 Effect of electron mediators on SRB influenced corrosion

MIC related parameters	Bacteria	Bacteria + Riboflavin	Bacteria + FAD
Corr. rate mm/year	0.012	0.028	0.022
Pit depth, μm	4.8	6.6	6
Pit diameter, μm	4.2	10.4	6.2
Residual SO_4^{2-} (mM)	1.7	0.8	1.2
SO_4^{2-} consumption*	0	53	29

*Increased (%)

This process of extracellular electron transfer (EET) is termed as mediated electron transfer (MET). Two other processes of EET have been discussed in an earlier chapter). According to BCSR theory, the electron mediators draw the electrons from iron surface by reducing themselves near the surface. These mediators ' Med(red)' then while passing near the cell cytochrome gets oxidized and transfer electrons inside the cell. The oxidized mediators Med(Ox) then travel toward the iron surface and undergo the same redox process (Fig. 2.1 lower part). These changes brought by the bacteria can be understood on electrochemical principle basis. Redox potential of riboflavin is cathodic to that of iron, so in a cell with the two electrodes the riboflavin act as cathode and results in its reduction and oxidation of iron. The equilibrium potential in this case will change to between the potentials of the two electrodes. This potential is anodic to the potential of cytoplasm. So when reduced riboflavin comes near cytoplasm, it gets oxidized, thereby transferring its electron to the cytoplasm which gets reduced. The reduction reaction SO_4^{2-} to HS^- takes place inside cytochrome. The rate of this process is much faster so as to explain high degree of MIC attack on 304 stainless steel in natural marine water. This is evident from the experimental results performed on SRB-induced corrosion of 304 stainless steel exposed to natural marine water in the presence of riboflavin and FAD as electron mediators. In the experiments, riboflavin and FAD were considered as electron mediators. 304 stainless steels was exposed to the anaerobic test solution ATCC1249 medium (approximating to seawater) inoculated with SRB with and without electron mediators. The test showed that (i) abiotic test solution with and without electron mediators showed average pit depth of 2.06 μm with no significant variation among the three cases. (ii) When SRB is present, the data are shown in Table 5.3.

The data clearly suggest increase in degree of SRB-induced corrosion due to addition of electron mediators. It is also evident from the table that increased corrosion is associated with higher degree of SO_4^{2-} reducing to sulfide.

5.10 Maritime Vehicles [15]

Most common biofouling sites are hulls of ships, where barnacles are often found. The most obvious problem of growth on a ship is the eventual MIC of the hulls, tanks and piping systems leading to the ship's deterioration. Even before corrosion occurs, if left unattended, organic growth can increase the roughness of the hull, thereby decreasing its maneuverability and increasing drag. This domino effect continues when the ship's fuel consumption increases, in some cases by 30%. This in turn has economic and environmental consequences, as increased fuel consumption leads to increased output of greenhouse gases. Economic losses are tremendous, as fuel accounts for up to 50% of marine transportation costs. The damage of these components can necessitate expensive repairs and in a worst case scenario can potentially lead to replacement of the vehicle.

While the environmental conditions onboard maritime vessels and the conditions in which they operate can vary significantly (e.g., merchant shipping versus naval ships), the temperatures, availability of nutrients, and oxygen levels are often suited for MIC related microorganisms. These are present in shipboard fuel tanks, bilges, engine, etc., and in nutrients containing waters of polluted harbors and ports (having nutrients for microbes). Additionally, the source of nutrients for them are cleaning products, fuel, and lubricants. The stagnant water in bilges, tanks, and pipes, particularly, when ships are docked for extended periods, generates anaerobic conditions. Possible mechanisms by which microbes observed in these environment influence corrosion could be formation of oxygen concentration cells, metabolic production of FeS, acids, e.g., sulfuric acid, etc.

Now, we take a look at some of the MIC cases observed on ships etc.

Around 1966, localized attack was reported in a ship's **bilge** causing deep pitting with perforation of 8 mm mild steel plate located near the propeller shaft casing in the ship's stern occurred in <2 years. Presence of ferrous sulfide, a metabolic product of the metabolism of SRB suggested the attack to be due to SRB related MIC.

Corrosion of **fuel storage tanks** and other fuel using equipment due to contamination of fuels has been observed to block purifiers and filters of navy ships. This has resulted in losses in fleet efficiency and has required expensive maintenance. The problem has been linked to SRB, fungi and yeasts. Similar problems have been reported due to microbial contamination of distillate fuel, at the water-fuel interface, causing damage of gas turbines necessitating expensive repairs. Lot of work has been conducted related to investigate and prevent production of toxic H_2S gas, a by-product of SRB growth in the shipboard environment. Testing of bilge water in engine room and other machine spaces of many naval vessels has observed the presence of SRB's, aerobic, and coliform bacteria. In some navy vessels, 10 mm bilge plate was penetrated in <1 year (equivalent to corrosion rate of at-least 10 mm/year). Problems in **gas turbine engines** due to MIC have been reported. The failure has been found to be linked to corrosion of cooling system which uses seawater as cooling medium. The seawater was found to have nutrients, polluted with marine organisms and it was stagnant for around 4 weeks. Another problem is about severe pitting (2 mm/year) of copper-nickel alloy tubing used in **cooling systems** of submarines. The problem was found to be related with use of polluted seawater sourced from non-tidal basin in place of clean seawater. Aerobic and anaerobic (SRB) bacteria were detected in these water. Another problem in steel **hull plate** in a ship's bilge was reported to have pits of depth 8 mm developed in 12 months. Analysis of bilge water and sludge revealed the presence of aerobic and anaerobic (SRB) bacteria, fungi, and yeasts. This appeared to be the case of MIC by SRB as corrosion products and mud samples detected the presence of sulfides. In a case of accelerated corrosion in the **ballast tanks** of a ship, perforations of stringers in less than 2 years (equivalent to corrosion rate of ~6 mm/year) were observed. It was suggested to be the case of MIC due to both SRB and aerobic bacteria.

In double hull crude oil tanker, MIC problem encountered was pitting of uncoated bottom plating at a rate of up to 2 mm/year. Testing showed MIC bacteria including SRB and acid producing bacteria in settled water and sludge at the bottom of cargo oil tanks and in water droplets in the crude oil. In another case, an 11 mm hull of a ship experienced MIC affected perforation in <6 months. It was attributed to the bilge waters of many ferries and tankers that showed widely spread aerobic and anaerobic SRB, yeast, and molds. One can imagine about the severity of MIC in the sense that corrosion rate in abiotic seawater with same chemistry was tested to lie within ~0.1 mm/year.

5.11   Offshore Infrastructure [16]

Among the infrastructures, those mainly related to oil production experience vagaries of MIC to the maximum extent. Although, corrosion in abiotic case in particularly anoxic water is of no technical consequence, but in case of anoxic environment having microbes, extreme corrosion is observed. Thus one notes immense level of corrosion in buried or subsea pipelines containing water and/or oil. It has been estimated that more than 20% of pipeline corrosion is attributed to microorganisms. This is due to the fact that offshore oil facilities provide vast, nutrient-rich environment for microorganisms. MIC frequently forms pits in carbon steel pipelines leading in extreme cases to pipeline failure and oil spills. Off shore facilities are particularly prone to MIC where reinjection of produced water, consisting of injected seawater and formation water, is practiced often. During reinjection, nutrients, e.g., dissolved hydrocarbons from oil and electron acceptors mainly sulfates from seawater as well as oxidants from the resident microorganisms provide necessary carbon and energy. Though oil industry uses deaerated water to reduce the corrosive effect of oxygen, the same practice creates anoxic environment in oilfield fluids where plenty of SRB and NRB (nitrate reducing bacteria) thrive. This condition enhances the reducing reaction rates, through oxidants, leading to enhanced corrosion.

There are two types of marine infrastructures (i) fixed structure are on the gulfs and bays, ship/rail yards, fuel terminal, platforms for oil drilling, etc., (ii) mobile structure are ships, yachts, patrol boats, submarines, etc. Infrastructure elements, located along the coastline, are made of steel, stainless steel, reinforced concrete, plastic and composites, elastomers for lining tanks and pipelines. There are two types of docks—a floating one and a dry one. The first is meant for inspection, maintenance, and repair jobs and incurs high maintenance cost due to frequent removal of corrosion products, sea fouling, and scale to minimize corrosion. It consists of ballast chambers by which the ship submerges in and emerges out from seawater depending upon the requirement. Due to this sort of wet/dry cycle, the floating dock experiences high degree of corrosion. When water in port is contaminated and the sea soil beneath the floating deck has corrosive sediments, accelerated corrosion results in perforations of the floating deck floor and seawater chambers. A dry dock is made of concrete, having steel gates, required while building up or repairing of the ship. After completion of the work, the dock is flooded and the ship slides into sea coast. Modern merchant and military ships built of steel are sailing the seas or moored at pier ports, therefore, they are affected by corrosion and fouling. Piers, and other aquatic structures are also affected by MIC. Problem area is splash zone just below low tide and microorganism responsible is sulfate-reducing bacteria (SRB). Offshore infrastructure also consists of a complex network of pipes, pipelines, and tanks. Biofilm forms on inner surfaces of most of these due to low velocity flow/storage of multiphase fluids which allows settlement of sediments and bacteria at the bottom of pipe and providing a favorable site for biofilm colonization. This accelerates corrosion and many of them experience pitting. Areas of locally reduced flow and stagnant flow conditions, e.g., heat affected zones of welds, pipe elbows, separators, valves, etc., and storage tanks are observed with enhanced corrosion.

Corrosion tests have been carried out in natural/artificial seawater. The diversity of the materials used in marine vehicles' building is governed by the requirement of the vehicles structural components, their function and the vehicles operating expectations. Thus, the materials considered for testing can be classified into three groups: (i) carbon steels (hull steel), low alloyed steels, and stainless steels, (ii) copper-based alloys, e.g., bronze and brasses, meant for vehicles' water supply and sanitary systems, air conditioning, heat exchangers, propellers, piping, etc., and (iii) aluminum-based alloys of series 5XXX, e.g., UNS A95052 for manufacturing navy, fast vessels, and recreation boats. The measurements were carried in several marine locations. Corrosion rates obtained by weight loss test are shown in Table 5.4.

One finds that carbon steel has poorest and stainless steel maximum corrosion resistance. In natural seawater, bronze, and aluminum show better corrosion resistance with aluminum the better among all except stainless steel. In synthetic seawater, copper is better than brass in terms of the corrosion resistance. At times when water becomes acidified due to the addition of acidified waste effluent, corrosion rates

Table 5.4 Corrosion rates in GOM seawater

Material	UNS	Type of Seawater	Location	CR (mm/year)
Carbon steel	G10450	Natural	L1	0.22
Stainless steel	S31600	Natural	L1	0.0033
Copper	C12200	Synthetic	L2	0.059
Bronze	C52100	Natural	L2	0.10
Brass	C26000	Synthetic	L2	0.20
Aluminum	A91050	Natural	L2	0.018

L1: Sound Campeche, L2: UV—Universidad Veracruzana, Anti corrosion unity have been observed to increase by ~4 times in case of carbon steel and ~2 times for stainless steel.

5.12 Organic Fuel [16, 17]

Organic fuels are the hydrocarbon fuels, e.g., gasoline, jet fuel/aviation kerosene, diesel, etc. These fuels are kept in storage tanks or may be present in fuel tanks of automobiles, airplanes, etc. Even in best kept tanks, microbial contamination is observed often. Microorganisms are usually present in fuel. For microbial growth, presence of water is essential and it is present nearly always in fuels due to (i) condensation of water, dissolved in fuel, on tank walls, (ii) entry of air moisture through tank lids or other openings, (iii) inefficient drainage of tanks, (iv) addition of water due to ship ballast. An amount of ~0.5 ml/l of fuel is enough for initial growth of microbes and 10 ml/l for substantial growth. Oxygen is normally present in sufficient quantities in distillate fuels and it is replenished at every refilling of tanks. However, even if the fuel becomes anaerobic, it is vulnerable to microbial attack due to facultative organisms, e.g., *Bacillus* and anaerobes SRB. Studies have shown that fungi grow readily in fuel system having mineral salts solutions as aqueous phase, which are present because of many additives used in fuel industry. MIC and bifouling, associated with microbial growth are usually observed in areas of fuel systems including storage tanks (Table 5.5). In general, most problem due to MIC is in storage tanks and pipework, formation of microbial mat/biofilm with the ability to block filters and pipelines and wear of pumps. Data suggest that as little as 1 mg particulates/100 ml fuel (~1%) can cause filtration problem.

Most aluminum alloys AA 2024 and 7075, affected by MIC, are used in aircraft or in underground fuel storage tanks. In aircraft fuel tanks, generally made of Al alloys, *Hormoconis resinae* utilizes the hydrocarbons of diesel fuel to produce organic acids. This leads to corrosion and penetration of tank linings. Surface in contact with aqueous phase of fuelwater mixture (observable in oil storage tanks) and sediments are common sights of attack. Organic acid by-products excreted by this fungus dissolve or chelate Cu, Zn, and Fe from the grain boundaries of aircraft aluminum alloy leading to inter granular attack and pits which persists under anaerobic condition under the covering of fungal mat. The impact of acidic metabolites is intensified

Table 5.5 MIC/biofouling observed in fuel systems

Problem	Principal microorganisms
Blockage of pipes, valves, filters	Fungi, polymer producing bacteria
Corrosion of storage tanks	Fungi and anaerobic bacteria
Breakdown of hydrocarbons	Fungi and aerobic bacteria
Penetration of protective tanks lining	Fungi

when trapped within metal-biofilm interface. Since fungi are present everywhere in atmospheric and aquatic environment, they accumulate organic material and produce organic acid, e.g., oxalic, lactic, formic, acetic, and citric acids. *Cladosporium resinae* (Kerosene fungus) in jet fuel, which grows in 80 mg of water per liter of kerosene one act to produce citric, cis-aconitic, and iso-citric acids. These acids induce pitting through decrease in pitting potential of Al alloys. Growth of fungi in diesel fuel storage tanks can produce large quantities of biomass and this may provoke crevice attack on the metal. *Fungal mycelia* binds metal ions, resulting in metal ion concentration cells on the metal surface. Differential aeration caused by the adhesive or fungal mats can cause crevice corrosion. Besides the decrease in bulk pH also accelerates corrosion due to metabolites produced during the growth by fungi.

Let us now take two case histories showing effect of corrosion in oil storage systems. In first case, increased wear and breakages were found occurring in vehicles of bus companies when a new type of additive was introduced in the diesel. Samples of fuel were collected from representative parts of the system for microbiological analysis. Most companies were not maintaining their diesel storage systems properly. A typical biomass was found in some storage tanks, accompanied by an aqueous phase of low pH (~3–5). Poorly maintained storage systems needed frequent changes of filters, both in the storage system and in the vehicles, and they observed reduced engine efficiency. Increased replacement of fuel nozzles was also noted, a problem linked to microbial contamination. The most significant organisms isolated were *H. resinae* and SRB, isolated from storage tanks, filters, and tanks containing filtered fuel. No SRB were found in the vehicle tanks or injectors, where no waterphase was present, but bacteria of the genus *Bacillus* were frequently encountered at all the points sampled. It was not possible to correlate the level of microbial contamination with premature failure of engine parts, but there was a relationship between the latter and the use of the new additive. In another case of one firm, considerable quantities of fuel was being lost through highly corroded in-ground storage tanks hence also causing pollution. The tanks were replaced by an aboveground system with storage tanks inclined at 30° to the horizontal to facilitate water drainage and removal, resulting in a better-maintained system and fewer problems.

References

1. Sanders PF (1988) Control of biocorrosion using laboratory and field assessments. Int Biodeterior 24:239–246
2. Little B, Wagner P, Mansfeld F (1991) Microbiologically induced corrosion of metals and alloys. Int Mater Rev 36:253–272
3. Cord-Ruwisch R, Kleinintz W, Widdel F (1987) Sulfate reducing bacteria and their activities in oil production. J Petroleum Tech 39:97–106
4. Al-Darbi MM, Muntasser ZM, Tango M, Islam MR (2002) Control of microbial corrosion using coatings and natural additives. Energy Sources 24:1009–1018
5. Muthukumar N, Rajsekar A, Ponmariappan S, Mohanan S, Maruthamuthu S, Murlidharan S, Subramanian P, Palaniswamy N, Raghavan M (2003) Microbiologically influenced corrosion. In: Petroleum product pipelines—a review. Ind J Exp Biol 41:1012–1022

6. Khouzani MK, Bahrami A, Hosseini-Abrari A, Khandouzi M, Taheri P (2019) Microbiologically influenced corrosion of a pipeline in a petrochemical plant. Metals 9:1–14
7. Licina GJ, Cubicciottti D (1989) Microbial induced corrosion in nuclear power plant materials. JOM 41:23–27
8. Suban M, Cvelbar R, Bundara B (2010) The impact of Stagnant water o the corrosion process in pipelines. Mater Technol 44:379–383
9. Clarke BH, Aguilera AM (2007) Microbiologically influenced corrosion in fire sprinkler systems. Autom Sprinkler Syst Handbook Suppl 3:955–964
10. Gavrilă L, Gavrilă D, Simion A-I (2003) Microbiologically induced corrosion and its mitigation. Cooling Cycl Sci Stud Res IV(1–2):127–135
11. Ashassi-Sorkhabi H, Moradi-Haghighi M, Zarrini G, Javaherdashti R (2012) Corrosion behaviour of Carbon Steel in the presence of two novel Iron oxidizing bacteria isolated from Sewage treatment Plants. Biodegradation 23:69–79
12. Stanaszek-Tomal E, Fiertak M (2016) Biological corrosion in the sewage system and the sewage treatment plant. In: Proceedings engineering: world multidisciplinary civil engineering-architecture-urban planning symposium, vol 161, pp 116–120
13. Zhang P, Xu D, Li Y, Yang K, Gu T (2015) Electron Mediators accelerate the microbiologically induced corrosion of 304 stainless steel by Desulfovibrio vulgaris biofilm. Bioelectrochemistry 101:14–21
14. Yan M, Zhang Y, Zhang R, Guan F, Hou B, Duan J (2019) Microbilogically influenced corrosion of marine steels within the interaction between steel and biofilms: a brief view. Appl Microbiol Biotech 11p. https://doi.org/10.1007/s00253-019-10184-8
15. Wade SA, Mart PL, Trueman AR (2011) Microbiologically influenced corrosion in maritime vessels. In: Proceedings of ACA symposium. "microbiologically influenced corrosion", vol 36, pp 68–79
16. Valdez B, Ramirez J, Eliezer A, Schorr M, Ramos R, Salinas R (2016) Corrsion assessment of Infrastructure assets in coastal seas. J Marine Eng Tech 15:124–134
17. Gaylarde CC, Bento FM, Kelly J (1999) Microbial contamination of stored hydrocarbon fuels and its control. Revisita De Microbilogia 30:1–10

Chapter 6
Mitigation of Microbial Induced Corrosion

Abstract After deliberating about basics related to corrosion, MIC, different types of bacteria, and their metabolic reactions taking part in the process of corrosion of metals, it is obvious to talk about the ways and means for mitigating MIC. This chapter deals with this aspect in two parts, the first part dealing with the conventional/currently undergoing techniques while the second part dealing with newer/green technology approach for MIC mitigation. One can refer it to MICI, i.e., microbial induced corrosion inhibition. Thus, it deals with not only inhibition of MIC but also microbial induced inhibition of chemical corrosion. First category of methods includes (i) cleaning, (ii) usage of biocides, (iii) coating, and (iv) cathodic protection. Conventionally, mostly adopted procedure to control MIC is by use of biocides. However, biocides pollute the environment, some of them are carcinogenic, and some even corrode material of construction. Consequently, newer methods being suggested have to be environment-friendly. Approach adopted for controlling MIC includes usage of microorganisms which affect (i) production of protective biofilm, (ii) in-situ production of antimicrobials, and (iii) usage of extracellular polymeric substances (EPSs).

Keywords Cleaning · Biocides · Coating · Protective biofilm · Antimicrobials · EPS

Until now in this book, we have discussed various aspects of MIC, e.g., what is MIC and how metals are affected by microbial corrosion, which microorganisms induce microbial corrosion and under which environment, what are the different industries affected by MIC and how it affects economy of production in different industries by attacking various materials of construction and thus affecting the maintenance, repair and replacement strategies, etc. Various agencies have conducted surveys dealing with the cost aspect incurred by the industry in different countries. Microbial corrosion has emerged out as a significant contributory of the total corrosion cost to the industry. In order to make any business viable, the owner/investor is always in the pursuit of searching for all possibilities of reducing the cost of operation so as to maximize profit. Mitigation of MIC is one such important point of consideration in this direction. All efforts are, therefore, tried in the industry to mitigate/control MIC.

The present chapter discusses the existing practices, the drawbacks associated with them, and the possibilities of adopting newer and better acceptable practices.

From the standpoint of application of MIC mitigation technique to industry, it should be more cost-effective to adopt preventive measures rather than paying for often difficult and expensive remediation treatments due to failures. Methods to prevent biocorrosion should address the following basic issues: (i) inhibition of the growth and/or metabolic activity of microorganisms and (ii) modification of environment, where corrosion is taking place, so that it avoids adaptation of microorganisms to the existing conditions, e.g., biofilm formation on a surface, etc. Procedures suitable for addressing these issues, so that MIC mitigation could be achieved, can be put in two groups—first the conventional/presently existing techniques and second the newer/green technology which intends to overcome the problems associated with techniques of corrosion mitigation of first group. These two groups of MIC mitigation have now been discussed in the following part of the chapter.

6.1   Conventional/Presently Existing Techniques of Biocorrosion Mitigation

These procedures can be put in five categories: (i) cleaning procedures, (ii) use of biocides, (iii) coatings, (iv) polymers, and (v) cathodic protection.

- Cleaning Procedure

'Keep the system clean' has been treated as the classical concept of mitigating MIC in industry. But practicing this method is very difficult. One should consider factors, as given in Table 6.1, before deciding about a cleaning procedure. The purpose of cleaning the system is to remove deposits formed on any surface, e.g., scaling and slimy sediments. **Scaling** is deposit of a hard substance, namely calcium carbonates, sulfates, or silicates due to precipitation of dissolved chemicals in aquatic media. The buildup of scale depends upon pH, temperature, water quality, concentration

Table 6.1 Deciding factors for selection of cleaning procedure

Adherence, hard or softness of material to be removed
Whether metal surface to be cleaned is hard, soft, brittle and reacts with chemicals which may be used in cleaning, etc.
Condition of the surface to be cleaned, e.g., smooth, rough, welded, etc.
Extent of cleaning required, e.g., partial, mirror finish, etc.
Limitations in cleaning due to geometry of structure, e.g., sharp corners, bends, slopes, planar or with steps, bumps, etc., and system components, e.g., inlet/outlet pipes, valves/pumps, monitoring equipment, etc.
Environmental impact, e.g., production of acidic/toxic gases, pollutants, etc.
Cost

of chemicals, and hydrodynamic conditions. Precipitation of calcium carbonate can be minimized by adding a mixture of different inorganic acids, e.g., HCl, H_2SO_4, or sulfamic acid. **Slimy deposits** (e.g., biofilm) are formed from suspended matter which adheres to surfaces. These deposits are mud, oil, bacterial slimes, etc. For cleaning slimy deposits, one uses 'flushing,' which is perhaps the most simple, although of limited efficacy. High fluid velocity (~1.5–2.5 m/s) may detach sediments, while lower velocity (<0.5 m/s) helps the formation of deposits. Mechanical methods used to remove/reduce biofouling should be used with filters of varying sizes depending upon the size of sediment particles to be removed. Care must be taken to choose selection of cleaning methods in case of a system with geometrical complexities. These inaccessible areas may remain incompletely cleaned which may become places of recontamination leading to localized corrosion due to differential aeration/concentration cells. Cleaning may include any of the following methods, e.g., brushing, pigging, cleaning spheres, or water jet. Some alternative tools for cleaning are cutters for evaporator tubes, rubber spheres for heat exchangers, mechanical pigs in production pipes or injection lines, and blasting with sand, grit, or water. Abrasive or non-abrasive sponge balls are frequently employed in industry. However, abrasive sponge balls can damage protective passive films which leads to accelerated uniform/localized corrosion, and non-abrasive sponge balls are not very effective with thick biofilms.

A special case is the use of flushing supported by cleaners or jointly with chemical agents that induce biofilm detachment. In general, a chemical cleaning is practiced following mechanical cleaning as former is more effective in dealing with areas of localized corrosion and remote zones. Chemicals useful for this method are (i) mineral acids, e.g., HCl, H_2SO_4, and sulfamic acid used after mixing corrosion inhibitors. Addition of inhibitor avoids the corrosion attack on pipe metal, e.g., carbon steel, stainless steel, etc., due to mineral acids. (ii) Organic acids including formic, acetic, and citric acids. They are weak acids, so less corrosive than mineral acids. Thus, there is no need to add inhibitors in them. These acids are used in systems which are incompatible with inhibitors. (iii) Chelating agents, e.g., EDTA or HEDTA. These are inorganic and organic compounds which form complexes with metal ions and thus effectively remove iron oxide, when used for cleaning steel pipe and copper oxide when used for cleaning brass/bronze pipes. Their efficacy is pH dependent. For more details on chemicals used as cleaning agent, the readers may refer to reference [1] and references therein. Acid cleaning is not applicable to welds on stainless steels unless they were previously heat-treated or solution-annealed. If stainless steel has not undergone this treatment, it will experience stress corrosion cracking near the welded parts, while in contact with acid, which may lead to its failure. Acid cleaning compounds are generally more selective in removing scale and corrosion deposits but are not so selective at removing biological deposits. In heavily fouled systems, mechanical cleaning should be done prior to chemical cleaning with scrapers, brushes, or balls. Thereafter, chemical cleaning treatments using synthetic polymers could effectively disperse fouling deposits. Polyacrylates, polymaleates, and copolymers of partially hydrolyzed polyacrylamides are commonly used as dispersant chemicals.

Table 6.2 Desirable properties expected in an industrial biocide

1	To what extent biocide is selective against targeted microorganism
2	Must not corrode the material being cleaned even in the presence of other chemicals and under the operating conditions of biocide usage.
3	In case biocide corrodes the material, advantages aimed to gain from use of biocide will become severely limited or may be just nullified
4	Biocides should be biodegradable so that its use does not pollute the media
5	Low cost, since overall its use should be cost-effective

- Biocides

Biocides are single or multiple compounds, oxidizing as well as non-oxidizing, which are used for terminating microorganisms or inhibiting their growth, e.g., chlorine, ozone, bromine, isothiozoles, glutaraldehyde, etc., with the aim to inhibit microbial corrosion. Biocides disinfect the system by inhibiting bacteria, fungi, and algae. A biocide which is meant for killing bacteria may not be useful for killing fungi or algae. Also, a biocide may be effective for a particular strain of a bacteria but not for other strains. Also to use a biocide, its optimal dose for effective action in a given system has to be decided. Table 6.2 depicts properties required in a given biocide.

Now one needs to be aware of the properties required in a biocide which should be considered, for biocidal action. Table 6.3 outlines these characteristics. Most common oxidizing biocides are chlorine, bromine, ozone, and hydrogen peroxide. While using them, one should take into consideration their side effects, namely (i) interaction with other chemicals present in the media, (ii) their tendency to corrode structural metals, and (iii) possible attack on nonmetals, e.g., plastic, rubber, wood. These effects should be evaluated considering their oxidizing power, dozes, and whether treatment is continuous or intermittent. **Chlorine** as biocide is fed into the system in gaseous form. But it dissolves in water and remains as HCl, HOCl (pH < 7), or OCl^- (pH > 9.5). Consequently, pH range of 6.5–7.5 (where form of chlorine is mostly HOCl and smaller amount of OCl^- ion) is considered ideal for biocidal action. Lower-level pH will enhance corrosion due to increase in acidity and Cl^- ion concentration while higher than 7.5 increases fraction of OCl^- ions which has poor biocidal action. The level of chlorine concentration is kept between 0.1 and 0.2 mg/l. However, care must be taken against decrease in chlorine concentration due to its tendency to penetrate bacterial biofilm. Measurements have shown chlorine concentration reduced to 20%, inside the biofilm, of the amount present in bulk liquid. As such, one observes decrease in chlorine biocidal efficacy with time while using it as biocide on a system having bacteria-containing liquid biofilm. Other chlorine salts that can be used as biocides having similar action as that of gaseous chlorine are sodium hypochlorite (NaOCl), calcium hypochlorite, and also chlorine dioxide gas. **Bromine** is more effective as biocide in a wider pH range as compared to chlorine. The compound of Br, i.e., hypobromous acid (HOBr), responsible for biocidal action, is more effective than hypochlorous acid (HOCl) up to a pH of 8.5. At this pH, the amount of hypobromous acid is 50% and the rest is OBr^- (hypobromite ion) whereas

Table 6.3 Important characteristics of biocides used in water system

Biocide	Effectiveness	Oxidizing/non-oxidizing	pH dependence	Usual concentration (mg/l)
Chlorine	Bacteria and algae	Oxidizing	Dependent	0.1–0.2 (continuous treatment)
Chlorine dioxide	Bacteria, less extent: fungi and algae	Oxidizing	Independent	0.1–0.2
Bromine	Bacteria and algae	Oxidizing	Wide pH range	0.05–0.1
Ozone	Bacteria and biofilm	Oxidizing	Dependent	0.2–0.5
Quats[a]	Bacteria and algae	Non-oxidizing	Surface activity	8–35
Glutaraldehyde	Bacteria, algae, fungi, and biofilm	Non-oxidizing	Wide pH range	10–70
THPS[b]	Bacteria, algae, and fungi	Low environmental toxicity	Specific action against SRB	–

[a]Quaternary ammonium compounds and [b]tetrakis hydroxymethyl phosphonium

in case of chlorine the amount of HOCl (responsible for biocidal action) is only 10%. **Ozone** has been considered as an alternative biocide in view of the recent concerns about environmental degradation due to using halogen-based compounds. It is more attractive to use as compared to other alternatives chemicals, for applying as biocide, due to (i) high oxidative power making it more effective against most of bacteria including biofilms as well, (ii) low degree of corrosivity toward majority of structural metals including mild steel, and (iii) antiscaling. Ozone concentrations of 0.2 mg/l are effective to control a system with low organic contamination. In general, the range of concentrations between 0.01 and 0.05 mg/l is sufficient to prevent biofilm formation. For surfaces covered with biological deposits, the concentration may be enhanced to 0.2–1.0 mg/l for removal of biofouling. **Non-oxidizing biocides** can be more effective because of their better overall control of bacteria, algae, and fungi. Also their action is pH independent. Frequently therefore, it is recommended to use a combination of oxidizing and non-oxidizing type of biocides for an optimized control of microbes. Examples are glutaraldehyde, THPS, acrolein, etc. THPS is a new and most promising among the non-oxidizing-type biocides. It is effective with wide spectrum efficiency on bacteria, algae, and fungi. It is widely used in oil industry due to its ability to dissolve FeS which helps in decreasing SRB-induced corrosion. Also it is less toxic environmentally. But these categories of biocides, by and large, are so toxic that they must comply with environmental regulation. Like oxidizing-type biocides, many of these compounds are very toxic and for this reason they have not been found acceptable to a large extent. More details about these biocides can be looked in available text (Refs. [1, 2] and references therein).

- Coating

This method is to provide over metal surface with a covering which prevents aggressive chemicals and microbes from coming into direct/indirect contact with metals. In this way, corrosion of metal is minimized or avoided and thus the covering protects the metal from chemical or microbial induced corrosion. Preferably, the material of the protective layer or so-called coating should be non-toxic. However, in order that coating is protective type against corrosion, it should be electrically non-conducting, compact so as to minimize diffusion of ions, adherent to substrate, i.e., metal in present case, continuous, without any defects, e.g., cracks or discontinuities. Coating with discontinuities leads to localized corrosion of metal. Corrosion-resistant coating, lining, or plating can be of corrosion-resistant material, e.g., stainless steel, titanium, etc., antifouling paints, plastic, ceramic, etc. The coating material should also be non-degradable by bacteria in the environment where it is used and it should not release corrosive products during degradation. On the basis of previous studies, coal tar and epoxy resin coatings have shown better performance while PVC-based coatings show poor performance. Cement lining although reduces microbial fouling, it may be attacked by SOB *Thiobacillus*. Recent work done on coating has considered (i) addition of natural additives in oil-based coating and (ii) electrodeposited Zn-Ni-chitosan coating.

Effectiveness of the oil-based coatings (alkyd) with natural additives has been tested to compare extent of biocorrosion on uncoated and coated mild steel in SRB (grown in lactate/sulfate culture medium) inoculated medium [3]. The test was done before and after adding natural additives, derived from olive oil and local fish oil, in the coating. Different metal coupons were placed in the SRB medium and kept in sealed containers, to ensure anaerobic medium, for 3 months. The corroded surfaces were analyzed by SEM. The uncoated mild steel showed bacteria colonies and biofilm attached to the surface and between the layers heavy and dense corrosion products. Steel coated with alkyd only showed bacterial colonies and biofilm. Breaching of the coating was detected on the surface, which led to severe localized corrosion. Black FeS deposits were also observed. Steel coated with alkyd mixed with olive oil did not show biofilms, but a few spots were observed at different locations on the surface. Blistering with and without rupturing of the coating was observed on the coated surface. Finally, steel coupon coated with an alkyd coating mixed with fish oil showed a few bacterial colonies and a very thin biofilm. No breaches, blistering, or deterioration were detected on the surface. The coated surface was found to be well protected. The study showed MIC attack proceeding at a lower rate on the surfaces coated with alkyd mixed with the natural additives with marked inhibition of bacterial adhesion and better corrosion resistance. Mixing fish oils with oil-based coatings like alkyd showed a positive result toward inhibiting both the biofilm formation and the MIC effects on mild steel surfaces.

Another work [4] demonstrated the effect of electrodeposited Zn-Ni-chitosan coating for MIC inhibition. Among all the Zn-group VIII elements (e.g., Fe, Co, Ni), alloy-based coatings, Zn-Ni coatings, have been found to be better in aquatic environment than others due to its high corrosion resistance and good mechanical properties providing an eco-friendly alternative to toxic Cd coatings. However, its inhibition performance was found to be unsatisfactory when applied in marine environment which is biotic due to MIC and biofouling. Particularly, one looked for development of an alloy which provides resistance against SRB-induced corrosion and biofouling in marine environment. It is already established that chitosan, through its cationic amino groups, interrupt the bacterial membrane and then disrupt the mass transport along the bacterial cells leading to death of bacteria. Accordingly, chitosan-containing Zn-Ni alloy was opted due to its green biocide character. This electrodeposited coating, with varying amount of chitosan, was subjected to 6-day long corrosion experiment for investigating corrosion resistance and antibacterial properties of the coating in media having SRB. It was observed that pH of the media increases from 6.5–6.7 to a maximum of 7.7–7.8 for coating with higher amount of chitosan. This change inhibited the metabolism and growth of SRB. This was also observed as lowest bacterial concentration observed in case of coating with higher amount of chitosan. Compared to the blank medium, the sterilization rate in case of these coating reached over 80%. Corrosion rate of metal with coating having no chitosan was observed to be 0.32 mm/year while that for coating with highest chitosan was 0.29 mm/year but for coating with 0.2 g/l chitosan corrosion rate was lowest, i.e., 0.27 mm/year. This was attributed to higher Ni content as compared to other chitosan-added coating. SEM images of coatings showed breakings and cracks in the coating with no chitosan but that showing lowest corrosion rate had intact and smooth surface. The important outcome was that corrosion resistance and antibacterial properties of coatings improved in all cases having chitosan as compared to the one with no chitosan.

- Polymers [5].

Biocide usage for mitigation of biocorrosion has been an effective technique for many years but of late, people have started realizing its disadvantages (discussed later in this chapter). Accordingly, the industries are looking for finding alternatives to usage of biocide. One such alternative and popular technique, which is benign, is to mitigate MIC by provision of a protective barrier between metal surface and the bacterial medium. The protective barrier could be an inorganic coating [6] which develops on some of the metals such as stainless steel, titanium, and its alloys when they are anodically polarized. This phenomenon is termed as 'passivity.' Another approach is by using an organic/polymer coating. Owing to their good barrier ability and better corrosion resistance, polymer coatings are considered as a better option for controlling biocorrosion. However, there are several shortcomings in adopting this approach: (i) Scratches/cracks/defects have been observed in coatings which acts as crevices which are responsible for colonization of bacteria resulting in localized MIC, and (ii) weak bonding between polymer coating and substrate leads to open space between them with moisture [5]. This causes growth of bacteria which in turn is responsible for development of anaerobic condition due to oxygen respiration by bacteria. This condition results in development of SRBs and localized corrosion because of them (iii) polymer coatings readily suffer from degradation by microbes since some of them use organic components in coating as their nutrients for growth. This is responsible for deterioration in barrier properties of the coating. To overcome the problem of microbial degradation, polymer coatings with biocidal functionality have been developed, in recent years, which inhibit attachment of cells and their growth on the coating. Polymers commonly used for preventing and controlling MIC consist of three categories: (i) traditional polymers incorporated with biocides, (ii) antibacterial polymers having quaternary ammonium compounds, and (iii) conductive polymers.

Traditional Polymers Incorporated with Biocides

These polymers are polyurethane (PU), fluorinated compounds, epoxy resins, coal tar epoxy, polyvinyl chloride (PVC), etc. Coal tar epoxy and epoxy resin coatings showed good performance against biocorrosion while PVC-based coating poor protection against MIC. Protection characteristic of polyamide epoxy as mid-coat and polymer as topcoat on mild steel exposed in natural and synthetic seawater was acceptable. The effectiveness in protecting steel against corrosion decreased from polyamide epoxy, PU, latex, and alkyd. For protection against biocorrosion, two requisites for polymer coatings are (i) no alteration of coating by bacterial attack and no release of corrosive products during degradation and (ii) decrease in bacterial adhesion and biofilm formation inhibition. To inhibit biofilm formation, the coatings are furnished with biocides with antibacterial character. PU is extensively used as coating due to their impermeability, good adherence to substrate, abrasion resistance, flexibility, and biocompatibility. To prevent their biodegradation, they are incorporated with biocide. For this purpose, 0.1–5% of non-toxic antibacterial agents are chemically, rather than physically introduced so they become part of PU coating and distribution of antibacterial agent is uniform throughout. Thus, incorporated PU film shows > 99% reduction in microorganisms and this property remains unaltered over time due to no volatilizing and leaching of antibacterial agents. Application of nanomaterials has also been tested in developing such coatings. Thus, there is an example of modification of the surface of graphene oxide (GO) nanosheets by grafting polyisocyanate (PI) chains. Corrosion tests involving salt spray test and electrochemical polarization test done on a PU sheet incorporated with 0.1% of PI-modified GO nanosheets showed significant improvement of the anticorrosion properties and ionic resistance of polymer coatings.

Antibacterial Polymers with Quaternary Ammonium Compounds (quats)

Quats, a broad category of cationic compounds, are extensively used as corrosion inhibitors (e.g., inhibitor against acid corrosion of iron and steel) and biocides. The synergistic effect between +ive charged quats ions and halide ions results in increased corrosion inhibition efficiency. This is credited to cohesive Van der Waals force between + head group/halide ion complex and positively charged metal surface (due to oxidation of metal). Quats anti-microbial action is due to its attack on cells resulting in dissolution of lipids and release of intracellular materials. Detergent-like properties of quats work against the formation of polysaccharides (which form matrix to construct biofilm which in turn protects bacteria) released during the process of bacterial colonization. Since quats act simultaneously on corrosion inhibition as well as antimicrobial agent, they can be used to inhibit biocorrosion by preventing formation of biofilm and through this annihilating bacterial concentration. Experiments done on 70/30 Cu-Ni alloy, with coatings of quats on it, show good bacterial inhibition efficiency and decreased corrosion rate in marine water with aerobic bacteria *Pseudomonas*. Further improvement in this coating system has resulted in improvement in antibacterial efficiency to 99% toward *Desulfovibrio desulfuricans* and reduction of corrosion attack by ~95% throughout the exposure period. More recently, antibacterial inorganic–organic hybrid coatings have been synthesized on the metallic substrate surfaces by a combination of layer-by-layer (LBL) self-assembly and surface-initiated ATRP (atom transfer radical polymerization) [7] or a consecutive surface-initiated ATRP [8] to enhance the compactness of the protective coatings and the corrosion resistance to biocorrosion.

Conducting Polymers

These polymers are conducting, environmentally stable having unique redox reactions. This makes them most promising polymers to replace chromate ion coating which is considered as hazardous environmentally. Among the conducting polymers, polypyrrole (PPy), polyaniline (PANI), and polythiophene (PBT) have been used as anticorrosion coatings on aluminum, mild steel, stainless steels, copper, and its alloys, e.g., brass, bronze, etc. PANI has also exhibited antibacterial characteristics against gram negative *Escherichia coli* and gram positive *Staphylococcus aureus*.

Conducting polymer microspheres of poly (*N*-methylaniline) (PNMA), synthesized from *N*-methylaniline by its oxidation polymerization, have shown inhibition of growth and proliferation of SRB together with reduction of corrosion rate of steel. This antibacterial property of PNMA in epoxy coating against SRB has also been demonstrated. Recently, a nitrogen-rich dual-layer coating of poly(4-vinylaniline)-polyaniline (i.e., PVAn-PANI) was developed on stainless steel by combination of surface-initiated ATRP and in-situ chemical oxidative graft polymerization [9]. The as-synthesized surface was further N-alkylated by bromohexane to generate poly-cationic PANI with antibacterial activity. The antibacterial and anticorrosion performances of the quaternized PVAn-PANI bilayers were ascertained by bactericidal array and electrochemical studies, respectively. The experimental results, performed on as-synthesized stainless steel, demonstrated not only high killing efficiency of bacterial attachment but also remained stable and highly resistant under the synergistic attack of aggressive anions (Cl^- and S^{2-}) and *Desulfovibrio desulfuricans*. In one of the latest developments [10], chemical synthesis of dedoped bromo-substituted polyaniline (Br-PANI) was achieved. Epoxy resin composite coatings containing 2.0 wt% of dedoped Br-PANIs (EBP) were also prepared [10]. The anticorrosion and antifouling performances of EBP coatings were characterized by accelerated immersion test, electrochemical impedance spectroscopy, XPS, antibacterial test, and field test. Results showed that EBP coatings presented an excellent protection after 100 days of immersion in a 12.0 wt% NaCl solution at 95 °C. Moreover, EBP coatings had a better antibacterial and antifouling performance than a pure epoxy coating and a dedoped PANI composite coating. Further, the anticorrosion and antifouling abilities of EBP improved with the increase of the bromine content. Conclusively, one can hope for successful biocorrosion control through antibacterial conductive polymer coatings thus offering a great potential for mitigating and preventing biocorrosion in the future.

- Cathodic protection

Corrosion mitigation by cathodic protection (CP) is a method where the system (substrate beneath the biofilm) to be protected against MIC is connected to another metal so as to make a galvanic cell whose cathode is the metal substrate (covered by biofilm) and the other electrode is anode. When the two electrodes are electrically connected, the anode metal corrodes while cathode metal remains protected. There are two types of CPs: First one is 'sacrificial anode CP' where the metal which is to act as anode termed as 'sacrificial anode' is chosen based on its anodic position in galvanic series, e.g., Mg, Zn, Al, etc., with respect to the metal of pipeline/system with biofilm which is to be protected. The other system (Fig. 6.1) is 'current-impressed CP' where the pipe to be protected is connected to -ive electrode of an external battery and other electrode to the +ive terminal of the battery. The cost of the system is governed by (i) the cost of anode material, costing a lot, which corrodes during cathodic protection process and needs to be replaced after sometime, and (ii) power requirement for protection. Thus, in one of the studies [11] related to cathodic protection of ships, for bare steel in wavy and rough seas the current requirement was 350 mA/m^2 while after polarization requirement was to maintain 100 mA/m^2, as long as the ship is on sea

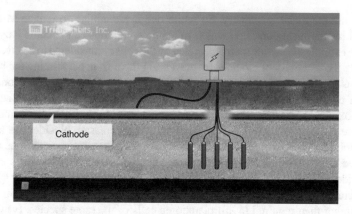

Fig. 6.1 Current-impressed cathodic protection of an underground pipe

(which could be weeks or months), and for coated steel ship, on wavy and rough seas, current requirement will be 50 mA/m^2 and after polarization it will be 15 mA/m^2. Thus, use of cathodic protection system on coated steel ship will be a cost-effective alternative.

Electrolytic systems have also been applied to prevent biofouling on ship hulls. In an underwater experiment, 36-m-long electrodes were fitted longitudinally to the bilge of a 126,000 ton ship under construction. This allowed uniform distribution of chlorine which acted as antimicrobial agent. Cathodic protection has also been used on ships, and a combination of cathodic protection and a coating is regarded as the most effective antifouling process.

6.2 Newer/Green Technology Approach of MIC/Biocorrosion Inhibition

Among the conventional/preexisting methods of MIC inhibition, most common is the use of biocides because it is comparatively easy to practice at industrial level too but there are limitations in using biocides. Other techniques like use of inhibitors and coating are also of limited use. Overall, there are drawbacks associated with these techniques of MIC/corrosion inhibition. Regarding application of biocides, there are several problems: (i) Unfortunately, they are inherently toxic and are difficult to degrade being persistent in natural environment or are able to accumulate in a variety of possibilities, e.g., water bodies with flow conditions, atmospheric gases, etc., resulting in contamination of pollutants even at places distant from the site of treatment. Overall biocides can have very negative impact on the environment and on the life of human beings, aquatic life, plants and crops, etc. This environmental concern has led to legislation which encourages replacement of biocides by environment-friendly approach. (ii) Oxidizing biocides, e.g., Cl$_2$, hypochlorite, etc., do corrode the metal machinery. So while one is attempting to reduce MIC, it is negated by chemical corrosion which in acidic conditions can become hazardous. (iii) The exopolysaccharide matrix secreted by the cells is considered to be involved in reducing the reactivity of biocides by preventing them from penetrating the biofilm. Attempts to circumvent this problem have focused on using very high concentrations of biocides which may in turn pose an environmental hazard. Consequently, neither oxidizing nor non-oxidizing biocides penetrate biofilms so bacteria, including SRB, are difficult to control within biofilms. For this reason, biocides are ineffective to inhibit SRB-induced corrosion which occurs in anaerobic conditions usually observed within/beneath the biofilms. Incidentally, SRB-induced MIC is also the most widely distributed and most severe type of corrosion attack on metallic systems. (iv) Bacteria inside biofilms have been observed to resist biocides at levels 500–5000 times higher than required to kill planktonic cells of the same species. (v) Continuous use of the same biocide makes bacteria resistant to the biocide or development of more bacteria resistant to the biocide being used. (vi) The working principle of biocides is based either on suppressing the growth and activity of bacteria from certain preselected groups or on the complete extermination of all microorganisms. A negative consequence of exterminating all bacteria is the removal also of those which have been reported to decrease the rate of metal corrosion. Problems associated with using coatings for inhibition of MIC are that they have been observed to experience microbial degradation. In case of coating with antimicrobials, cases have been reported of dissolution of antimicrobials from the coating into the environment which raises serious concerns. Another method of MIC inhibition is cathodic protection method which uses a sacrificial anode, a metal which preferentially corrodes being anodic to the protected metal. This method, however, turns out to be expensive and only works in certain environments. Moreover, it is a general observation that conventional methods of biofilm control and removal in an industrial setting are inadequate against the adherent biofilm bacteria.

Due to the shortcomings in adopting conventional techniques outlined above, inhibition of corrosion/microbial induced corrosion (MICI—microbial induced corrosion inhibition) through microbial route is being seriously considered as an important green technology approach alongside other options instead of biocides or other measures for MICI.

Microorganisms can contribute to corrosion inhibition by different mechanisms [12]. The main mechanisms of bacterial corrosion inhibition are always linked to a marked modification of the environmental conditions at the metal–solution interface from corrosive to metal-friendly environment due to biological activity. Further, (i) MIC and its counter-process, microbial inhibition of corrosion, are rarely linked to a single mechanism of a single species of microorganisms. Either a complex biofilm or protective film forms on a metallic surface inside which the corrosive or the inhibitory action of a bacteria develops, respectively. (ii) Accordingly, microbial induced corrosion and its inhibition microbiologically, in a metal aqueous system, can be observed in the presence of at least two bacteria or could be two species of a bacteria. One of them is contributing to corrosion and other its inhibition. Three aspects related to MICI are: (i) neutralizing the action of corrosive substances present in the environment, (ii) forming protective films or stabilizing a preexisting protective film on a metal, and (iii) inducing a decrease in the medium corrosiveness. As such, microbial corrosion inhibition is frequently accomplished through: (i) a decrease in the cathodic rate by microbial consumption of a cathodic reactant (e.g., oxygen consumption by respiratory activity); (ii) decreasing the medium aggressiveness in restricted areas of the metal solution interface, e.g., by neutralizing acidity, growth inhibition of corrosion-causing bacteria or from antimicrobial production by non-corrosive microorganisms, and (iii) providing or stabilizing protective films on the metal (e.g., biofilm exopolymers with metal-binding capacity). In practical situations, the inhibitory action of bacteria can be reversed to a corrosive action within bacterial consortia structured in the biofilm thickness. Finally, a proper understanding of the identity and role of microbial contaminants in the specific environment of a metal surface may be exploited to induce corrosion inhibition by bacteria as a useful tool to prevent frequent MIC effects encountered in practice. On the basis of work done so far, MICI can be practiced following options as per Table 6.4.

Besides there is also some work on MICI related to (i) genetically engineered microorganisms, e.g., *Bacillus subtilis, Bacillus licheniformis, E. coli*, and *Beauveria bassiana*, (ii) some other but unidentified bacteria, e.g., *Bacillus subtilis* BE1500 and *B. subtilis* WB600 on aluminum, mild steel, and brass, (iii) corrosion inhibition due to microbial induced carbonate precipitation (MICP) on ceramics, e.g., calcite, silicate, concrete, mortar, and limestone. A brief about some of the work is given below.

- Oxygen Removal (Protective Biofilm) [13]

It is known that Pseudomonas (and other aerobic bacteria) show both increase and decrease in rate of metal corrosion by way of producing polymer (EPS—extracellular polymeric substance) during biofilm formation on metal. Two more observations have been made in this regard: (i) Single culture of aerobic *Bacillus* sp. induces greater corrosion on mild steel initially, but the rate goes on decreasing with time and reaching equal to that of the sterile medium in ~17 days, and (ii) it also generates anaerobic medium in the metal–biofilm interface region. Some work has also suggested that under certain conditions Pseudomonas have controlled corrosion by preventing diffusion of corrosive substances through the biofilm. The biofilm appears to act as diffusional barrier. But there are different thoughts on whether the live/dead biofilm, formed in these circumstances, acts as a diffusional barrier or decrease in the concentration of oxygen at metal–biofilm interface happens due to respiration of oxygen by bacteria (through oxygen reduction reaction). To understand this aspect of biofilm, experiment with an axenic culture for aerobic bacteria was conducted in seawater-mimicking medium to simulate a marine environment. The purpose was to know the mechanism of corrosion taking place at metal–biofilm interface. With its help, one can understand how to inhibit colonization of SRB at biofilm (due to aerobic bacteria)—metal interface (this happens because of generation of anaerobic medium at the biofilm—metal interface) so as to avoid localized corrosion. Weight loss and EIS measurements were conducted to observe corrosion rate of steel sample in sterile aerobic and anaerobic distilled water, three (LB, VNSS, and Barr's media) aerobic and anaerobic media inoculated with *E. Coli.* and *P. fragie K* bacteria. The

Table 6.4 Overview of MICI studies using different types of organism and materials

Microorganism		Engineering material	Mode of action
Type	Species		
Bacteria	*Pseudomonas* spp. and *Escherichia coli*	Mild steel 1018	Protective biofilm (oxygen removal?)
	Pseudomonas Sp9 and *Serratia marcescens*	Steel	Oxygen removal
	Bacillus brevis	Mild steel 1018	SRB inhibition by production of antimicrobials
	Pseudomonas *K fragi* and *Bacillus brevis* 18	Unalloyed copper and aluminum alloy 2024	Protective biofilm (oxygen removal?)
	Pseudomonas cichorii	Mild steel	Protective biofilm and phosphate precipitation layer
	Shewanella algae and *S. ana*	Aluminum, mild steel, and brass	Protective biofilm
	Shewanella oneidensis strain MR-1	Mild steel 1018	Protective precipitation layer
	Pseudomonas flava	Mild steel	Protective biofilm and phosphate precipitation layer
	Bacillus licheniformis and *B. subtilis*	Aluminum 2024-T3 and C26000 brass	Production of anionic corrosion inhibitors
Fungi	*Aspergillus niger*, *Aspergillus alliaceae*, *Penicillium* sp., *Beauveria bassiana*, and *Fusarium* sp.	Bronze and copper	Protective oxalate layer

SRB sulfate-reducing bacteria
References for the work cited in Table 6.4 can be sought from Ref. [12]

results indicated that (i) both bacteria inhibited corrosion 3–6 fold compared to sterile controls (all media aerobic) and corrosion, in inoculated media, was comparable to that in anaerobic sterile media. This basically demonstrated that decrease in corrosion rate was due to inhibition of diffusion of oxygen by the biofilms formed of the bacteria which led to depletion of oxygen at the biofilm—metal interface. (ii) The decrease in corrosion rate was similar for all the three media inoculated by both the bacteria individually. This shows that inhibition of corrosion was by and large independent of metabolic products formed by the two bacteria in the three media. (iii) A very important finding of this experiment was the demonstration that live biofilms (or the presence of live bacteria) are necessary to prevent passage of oxygen through biofilm. It was concluded from the observation that open-circuit potential of steel in media decreased (with respect to that in sterile aerobic media) only as long as bacteria was alive but increased when bacteria was dead. Conclusively, therefore, these experiments established that the 'presence of live biofilm (or live respiring cell) is necessary to act as protective film to inhibit corrosion of steel induced by axenic culture of aerobic bacteria through oxygen depletion.' To further establish the mechanism of protection and their potential to protect by biofilm of *P. fragi* and *E. coli* DH5α in LB medium on mild steel, the experiments, as above, were carried in temperature varying from 23 to 30 °C so that observation could be made in the biofilm's changed metabolic state. Also the effect of addition of antibiotic kanamycin to kill biofilm bacteria in situ was studied to see their role on corrosion inhibition [14]. At 23 and 30 °C, in a rich medium, *P. fragi* and *E. coli* protect carbon steel by inhibiting corrosion 2–10 fold over 4 weeks of exposure. In this study, *P. fragi* was more efficient than *E. coli* in inhibiting corrosion at 23 °C, whereas *E. coli* was more efficient at 30 °C. To determine the protective ability of a dead (non-viable) biofilm, it is essential not to disturb the biofilm. Therefore, the cells were killed in situ, with minimal disruption of the biofilm, by the addition of the kanamycin. Dead *P. fragi* and *E. coli* biofilms did not inhibit corrosion as effectively as a live biofilm and allowed corrosion to proceed at rates comparable to the sterile control once the film was killed. This establishes that live biofilm (live respiring cell) is essentially required for inhibition of corrosion. When cells were transferred to a rotary shaker without an initial stationary phase, virtually no protection was observed, though comparable cell numbers were obtained in the supernatant. When the flasks were transferred after a weeklong stationary incubation period, the *P. fragi* K biofilm exhibited corrosion inhibition and it was nearly fivefold less than that in stationary flasks. This observation indicates at the importance of stationary incubation period for a bacteria to protect steel against corrosion though its biofilm.

- Protective film of axenic culture of *P. fragi* and *Bacillus brevis* [15].

Toxicity of copper to microorganisms has led to a notion that copper will not corrode significantly by microbial induced corrosion. However, metabolic products of some bacteria, e.g., ammonia by nitrate-/nitrite-reducing bacteria (NRBs), sulfuric acid by *Thiobacillus*, and H_2S by SRB, can cause corrosion of copper alloys. It has been suggested that the presence of a biofilm on copper creates differential aeration cells and chloride gradients, which can cause pitting. Thus, MIC of copper alloys is a problem in heat exchanger tubing, ship seawater piping, and aircraft fuel tanks. It has been observed [16] that the corrosion of copper in freshwater and seawater was inhibited by the addition of bacteria but corrosion increased after bacteria died, something recalling the necessity of the presence of live bacteria (or biofilms) for inhibition of MIC. Aluminum is considered to have good corrosion resistance due to formation of its passive oxide layer. However, *Pseudomonas* and *Cladosporium* species induce microbial corrosion on aluminum and its alloys. Pitting of these alloys and 100 times weight loss observed in SRB inoculated as compared to the extent of corrosion in sterile media has been reported.

In the pursuit of establishing the protective character of biofilm, another set of experiments were performed to investigate protection of Cu and Al alloy in modified Baar's medium inoculated with axenic culture of *Pseudomonas fragi* K and *Bacillus brevis* 18. In the experiments, unalloyed copper and aluminum alloy 2024 plates were tested. EIS measurements were done in sterile as well as inoculated Baar's media. Results of the test are given below.

According to Table 6.5, one can infer (i) higher polarization resistance (4.3-fold) in case of aluminum alloy with *P. fragi* K and sevenfold with *B. brevis* biofilm shows inhibition of MIC in comparison with the case of sterile media. In case of unalloyed copper, this increase of polarization resistance was ~20-fold with *P. fragi* K biofilm and ~ 2000-fold with *B. brevis* 18 biofilm. This is also supported by lesser anodic value of E_{corr} in case of metals with biofilm as compared to metals exposed to sterile media. (ii) It was ensured, by EIS experiments performed with nutrient flow, that inhibition of MIC in the two cases was due to biofilm and not due to planktonic (suspended) cells. (iii) To evaluate the effect on protective property of biofilm during SRB attack, SRB was added alongside any of either *P. fragi* or *B. brevis* MIC. One observes decrease in R_p and more anodic E_{corr} value indicating increased corrosion due to SRB. Another support of increased corrosion due to SRB growth was strong smell of H_2S which forms due to reactions responsible for SRB-induced corrosion. This work establishes that (i) Al and Cu experience MIC, e.g., by SRB, although in many sterile conditions they are resistant to corrosion. (ii) Addition of *P. fragi* and *Bacillus brevis* inhibits generalized corrosion on both by forming a protective biofilm.

- MIC inhibition by in-situ production of antimicrobials

Another method of MIC inhibition using microbes could be through in-situ production of antimicrobials. One such study relates to inhibiting MIC by inhibiting SRBs in biofilms on steel through in-situ production of antimicrobial peptides [17]. It was earlier believed that MIC due to SRB cannot be controlled without using biocides. Further, SRBs are inherently resistant to antimicrobials, e.g., ampicillin, chloramphenicol, etc., and not beyond than what is observed for planktonic cells. Further, the environment in which SRBs thrive (created by corrosion products) reduces antimicrobial efficiency. This is so because once SRBs are settled in biofilm, it is very difficult to eliminate them without removing the biofilm itself. However, it is observed that addition of antimicrobial prior to SRB colonization has shown complete inhibition

Table 6.5 Polarization resistance (R_p) and corrosion potential (E_{corr}) of aluminum 2024 and unalloyed Cu in modified Baar's medium with *Pseudomonas* and *Bacillus biofilm*

Biofilm	R_p (Ω cm^2)	E_{corr} (mV vs. Ag/AgCl)	Metal
Sterile	3.0×10^4	−670	Aluminum 2024
P. fragi K	13.2×10^4	−520	Aluminum 2024
B. brevis 18	21.3×10^4	−512	Aluminum 2024
Sterile	470 ± 220	−171	Copper
P. fragi K	9160 ± 2200	−118	Copper
B. brevis 18	9.66×10^5	−177	Copper
B. brevis 18 + SRB	1.43×10^5	−385	Copper

Table 6.6 Corrosion loss of SAE 1018 steel in modified Baar's media added with aerobic bacteria *P. fragi* K and *D. vulgaris* (SRB) in the order from left to right

Case	Medium	10 days	14 days	21 days
1	Sterile	0.77	0.88	1.03
2	*P. fragi* K	0.19	0.33	0.38
3	*P. fragi* K + *D. vulgaris*	0.35	0.52	0.71
4	*P. fragi* K + *D. vulgaris* + ampicillin	0.35	0.42	0.56
5	*P. fragi* K + ampicillin + *D. vulgaris*	0.25	0.33	0.49
6	*D. vulgaris*	0.19	1.23	~2.5

Gramicidin producing, *D. vulgaris*, *B. subtilis*, and *P. fragi* K is ampicillin and gramicidin resistant

on stainless steel/or delayed growth of SRB on steel. Inhibition effect, on both steel and stainless steel, has also been observed by externally adding peptide antimicrobial agent Gramicidin S prior to addition of SRB. As a novel strategy, it was therefore tried to inhibit MIC by addition of *Bacillus brevis* Nagono strain which secrets Gramicidin S and inhibits SRB on stainless steel.

Addition of *P. fragi* K in modified Baar's media (case 2), the mass loss of steel is observed to decrease with respect to that in sterile medium (case 1) (Table 6.6). This is due to formation of protective biofilm which minimizes the diffusion of oxygen toward metal–biofilm interface. Whenever D. vulgaris is present in the biofilm (case 6), the coupons were covered with a thick, black deposit due to sulfide formation and were difficult to clean. In this case, mass loss due to SRB-induced corrosion becomes higher than that observed in sterile medium due to colonization of SRB cells over metal surface with time. A dual culture of *P. fragi K* and *D. vulgaris* produced a 1.8-fold increase in corrosion after 21 days of exposure compared to a monoculture of *P. fragi* K (0.71 and 0.38 in case 3 and 2, respectively); however, the corrosion observed in both cases was always lower than that observed with sterile-modified Baar's medium. The corrosion loss observed with a monoculture of *D. vulgaris* on SAE 1018 steel (case 6) was higher than that in sterile medium after 14 days (1.4-fold, 1.23) and 21 days (2.5-fold). Consider now when *P. fragi* and *D. vulgaris* are added so that *D. vulgaris* is permitted to colonize the metal coupon and afterward ampicillin (100 μg/ml) is added (case 4). The mass loss of the coupon due to addition of ampicillin, in this case with respect to that where ampicillin is added prior to addition of *D. vulgaris* (case 5), is observed as 29% ([1–0.25/0.35 = 0.29] or 29%) more in 10 days to 13% ([1–0.49/0.56 = 0.87] or 13% more in 3 weeks. This indicates that addition of ampicillin in medium inhibits SRB-induced corrosion with the condition that SRB should not be allowed to colonize metal surface before addition of ampicillin (Table 6.6).

EIS measurements further confirmed above observations. Thus, growth of *D. vulgaris* in the reactors increased R_p 90-fold after 72 h compared to sterile controls. This increase inR_p or decrease in corrosion is basically due to (i) much lesser growth of SRB due to media being aerobic, and (ii) the change appears after some time when SRB cells have colonized over metal surface. Addition of 200 μg/ml ampicillin

Table 6.7 R_p (Ω cm^2) values derived from EIS measurements on 304 stainless steel

Case	Media	R_p (Ω cm^2)
1	*P. fragi* K and *P. fragi* K + ampicillin + SRB	2×10^7
2	*P. fragi* + ampicillin + SRB	1.9×10^7
3	*P. fragi* + gramicidin S + SRB	1.44×10^7
4	*B. brevis* 18 + SRB	1.69×10^7
5	*P. fragi* + SRB + ampicillin	4.6×10^5

after 240 h of SRB growth did not change R_p, since metal surface has already been colonized by SRB and there is no effect on resulting biofilm by ampicillin. The reactor remained black with the distinct odor of sulfide. The addition of *D. vulgaris* to a continuous *P. fragi* K reactor increased the R_p of mild steel threefold after 36 h, the reactor turned black, and the odor of sulfide was detected from the reactor outlet. Here, the increase in R_p or decrease in corrosion rate (with respect to sterile medium) is lesser (cases 1 and 3 of Table 6.7) as compared to that in case where *D vulgaris* is added in sterile water (case 2 of Table 6.7) due to addition of SRB which enhances corrosion resulting in the formation of sulfide. Addition of 200 µg/ml ampicillin and a combination of 100 µg/ml ampicillin and 25 µg/ml chloramphenicol after 120 and 150 h of SRB growth also did not shift R_p to its value prior to SRB addition, indicating that there was no inhibition of SRB cells after they have colonized on mild steel surface. The stainless steel EIS spectra for *P. fragi* K and *P. fragi* K + ampicillin + SRB showed high R_p values close to 2×10^7 Ω cm^2 (cases 1 and 2) indicating uniform corrosion typical of stainless steel in neutral media. From the values of R_p (Table 6.7) for different media, one can see there is practically no change in R_p for different cases (1–3) where antimicrobials are added prior to adding SRB and case 4 where gramicidin S is produced in situ but a significant reduction in R_p (~38 times) and corresponding increase in corrosion of stainless steel when SRB is added prior to antimicrobial (case 5). Considering approximate equivalent impedance circuits and fitting the experimental data to these circuits, it appears the corrosion rate of 304 stainless steel was reduced by approximately 50-fold by adding ampicillin or gramicidin S prior to SRB and by producing gramicidin S in situ using *B. brevis* 18 before adding SRB. Further, the absence of changes in the impedance spectra (before and after SRB addition), in cases 1–4, demonstrates inhibition of SRB on stainless steel when purified antimicrobials were present prior to the addition of SRB or when gramicidin S was generated by the biofilm itself. This study helps us in considering another green technology approach for inhibiting microbial corrosion through antimicrobials producing microbes. Since biofilms can form rapidly on exposed surfaces, this system can exclude other bacterial species from the biofilm. Such a system also has the added advantage of being able to reduce the extent of generalized corrosion caused by oxygen by as much as 40-fold. Thus, production of antimicrobial peptides within a protective biofilm to inhibit the growth of SRB is an attractive alternative to the use of high biocide concentrations. In addition, the approach discussed here might be appropriate for medical applications, for example, inhibiting the colonization of dental implants by deleterious bacteria which cause infection and subsequent implant failure. It may be true not just in case of dental implants but also in cases of implants anywhere in the human body.

- MICI using EPS

This method proposes to make use of layer of extracellular polymeric substances (EPSs) for MICI since an isolated and purified EPS has been shown experimentally to be protective in nature. However, to have a layer of EPS over a surface, most common way is to get deposited biofilm of bacteria. But biofilm is a matrix of EPS which keeps sessile cells embedded in it. So even though EPS may be protective type, the presence of cells will induce corrosion attack on metal–biofilm interface leading to MIC of metal. However, it has also been shown that application of EPS from *Desulfovibrio alaskensis* inhibited attachment of corrosive species Desulfovibrio indonesiensis and thus the formation of biofilm. Similar effects have also been observed with EPS from *Lactobacillus helveticus* which inhibited the attachment of *Listeria*. Since protective nature of biofilm results from EPS, efforts, in recent past, have been made to design practical experiments for applying EPS on metal surface and test their property of corrosion inhibition on metal surface [18–20]. In the first attempt [18], different strains of *Desulfovibrio*, namely *D. vulgaris*, *D. Indonesiensis*, and *D. alaskensis* were cultured in appropriate medium. The EPS of these strains, which were to be evaluated, was separated from bacteria by centrifugation and filtering. The filtered supernatant was dialyzed and freeze-dried in an evacuated desiccator. For testing their corrosiveness, aliquot of EPS suspension were deposited on metal coupons. Commercially available EPSs, e.g., alginate, dextran, and xanthan, were also tested. Degree of corrosivity of all these EPS in aerobic as well as anaerobic conditions on pure iron, carbon steel, and alloyed steel samples is shown in Table 6.8. Thus, one sees: (i) aerobic media. EPS from most strains, except Dextran, is corrosive to pure iron and carbon steel, which can be attributed to oxygen. On stainless steel, they are non-corrosive. (ii) Anaerobic media. EPS from *D. vulgaris*, *D. Indonesiensis*,

Table 6.8 Corrosivity of EPS on pure iron, carbon steel, and alloyed steel in aerobic and anaerobic media

EPS from strain	Metal					
	Aerobic			Anaerobic		
	1	2	3	1	2	3
D. vulgaris	C	C	NC	C	C	NC
D. Indonesiensis	C	C	NC	C	C	NC
D. alaskensis	C	C	NC	NC	NC	NC
Alginate	C	C	NC	llC	llC	NC
Dextran	NC	C	NC	NC	C	NC
Xanthan	C	C	NC	llC	llC	NC

Pure iron—1, carbon steel—2, alloyed steel—3
C—corrosive, NC—non-corrosive, llC—lot less corrosive

alginate, and xanthan is corrosive and lot less corrosive. *D. alaskensis* and dextran are non-corrosive on pure iron and carbon steel. They are all non-corrosive on stainless steel. This program was further extended to test EPS, including those of more bacteria, by coating them on metal samples to see its effectiveness in protecting steel against MIC [19]. For this study, various strains of bacteria including SRB and Pseudomonas were cultivated and their EPS harvested. These substances were tested for their inherent corrosiveness toward three most useful metals useful in industry, namely pure iron, carbon steel, and 304 stainless steel. Strains of bacteria tested in this study are outlined in Table 6.9.

Electrochemical polarization tests were done on all three metal samples, under aerobic conditions, and were taken with metal samples dipped in EPS dissolved in 0.2 M K_2SO_4 solution. E_{corr} was first monitored for 1 h, which ranged between -700 and -650 mV (w.r.t. sce) after which polarization curves were recorded to knowR_p (Table 6.10), from which corrosion rates were calculated.

Current potential curves were recorded for all the three metals in the presence of EPS harvested from various bacteria strains and commercial EPS as given in Table 6.10. EPS of *Desulfovibrio* strains does not show any effect on oxygen reduction, because these are anaerobic bacteria where oxygen does not play any role in corrosion. However, EPS of two Pseudomonas strains No. 4 and 5 and 8 and 9 (Table 6.9) shows inhibitive effect on oxygen reduction as evidence from concentration polarization (oxygen diffusion) part of cathodic curve. This occurs due to availability of much lesser amount of oxidant, e.g., oxygen in this case. The inhibitive effect of all EPS is also observed in the form of shifting of potential for onset of anodic oxidation in these cases by about 60 mV. Intrinsic corrosiveness of EPS on pure iron and carbon steel was observed under the anaerobic condition in case of EPS of *Desulfovibrio vulgaris* and *Desulfovibrio indonesiensis* and no corrosion (Table 6.8) by the EPS of *Desulfovibrio alaskensis*. From this standpoint, *Desulfovibrio alaskensis*—EPS appears to be a potential candidate corrosion inhibition by biopolymers.

The electrochemical behavior of the metals, in aerobic media, in the presence of EPS of *P. cichorii*, *P. flava*, *R. Opacus*, and *L. fermentum* shows inhibitive effect on oxygen reduction. The onset of anodic current, a representation of oxidation of metal, at higher potential in the presence of EPS, is an indication of inhibitive action by EPS on biocorrosion.

Another approach to inhibit MIC by using EPS is to investigate its effectiveness in resisting the colonization of SRB cells on the metal surface. One promising route can be to inhibit the adhesion of single cells. It is established that the processes of adhesion and desorption of microorganisms are induced and mediated by various biomolecules. It is believed that biofilms of bacteria have these molecules. Accordingly, the study involved harvesting EPS of various bacteria and purify them. Afterward, these EPS were adsorbed on metal substrate in order to form their layers which hopefully inhibit adhesion of cells on EPS-covered metal substrate. Guided by these ideas, efforts have been made to deposit adsorbed layer of EPS of *Desulfovibrio vulgaris* over metal substrate. Examination of these EPS-covered surfaces by epifluorescence microscopy and atomic force microscopy showed that the number of attached cells was significantly lower on the covered surfaces when compared to

Table 6.9 Strains of different bacteria for deriving their respective EPS

S. No.	Strains of bacteria
1	*Desulfovibrio indonesiensis*
2	*Desulfovibrio vulgaris*
3	*Desulfovibrio alaskensis*
4	*Pseudomonas flava*
5	*Pseudomonas cichorii*
6	*Pseudomonas putida*
7	*Pseudomonas fluorescens*[a]
8	*Rhodococcus opacus*
9	*Lactobacillus fermentum*
10	*Lactobacillus acidophilus*
11	*Citrobacter freundii*
12	*Enterobacter aerogenes*
13	*Arthobacter* sp.
14	*Xanthan*[b]
15	*Alginate*[b]

[a]Two strains of *Pseudomonas fluorescens* were of different origins
[b]Commercial extracellular polymeric substances (EPS)

Table 6.10 Corrosion-related parameters from electrochemical polarization curves

	Pure iron	Carbon steel	304 stainless steel
R_p (Ω) (Corrosion rate, mm/year)	95–150 (0.53–0.33)	78–133 (0.66–0.38)	Insignificant

pure substrates. Additionally, most of the EPS were found to be resistant against degradation by *Desulfovibrio vulgaris* [20].

Additionally, there are other options also available or in the process of development by which MICI can be attained using biotechniques through metabolic reactions/products (Table 6.4). Basic approach suggested is based on the fact that the bacteria which cause corrosion inhibition are by removal of oxygen or any oxidant. This leads to drop in cathodic reaction rate and slows down metal oxidation. The protective bacteria act as anode and metal as cathode in the galvanic couple. An attempt has been made by using a random phage [21] display library to look for novel aluminum and mild steel-binding peptides that have the potential for protecting these metals against corrosion. After the formation of biofilm, the bacteria can remain stable on metal surface and the expression of metal-binding peptides on the surface of such biofilm-forming bacteria may enhance the stability of organic coating and the efficiency of protection afforded but the peptides. Although some work has been put in this direction, still it is worth putting more efforts. Another novel SRB corrosion control approach is biocompetitive exclusion (BE) where corrosion-causing bacteria, e.g., SRB, are excluded from local microbial colonies by promoting the growth of competitive bacteria, e.g., nitrate-reducing bacteria. The dominance of beneficial biofilm helps in reducing corrosion attack [22]. Now let us check the effectiveness of taking help of laboratory results for utilizing them for finding a novel strategy, on their basis, of corrosion inhibition in industrial applications. One has to consider the fact that (i) most laboratory-based studies utilize glass (microscopic cover slip) or polymer (petridish, etc.) as substrate for biofilms instead of metals, (ii) tests are conducted in bacteria-rich medium favoring biofilm formation for using the knowledge gained by these experiments, and (iii) usually tests are performed in single bacterial regime not multiple one so that results can be analyzed easily and without much error. These conditions in general do not exist in industry; e.g., (i) the knowledge generated from the laboratory experiments is to be utilized for protecting the industrial machinery from MIC/corrosion. And most of the machinery is constructed of metals, so one needs to know the effect of biocorrosion on metal and for this the substrate needed for biofilm should be a metal, e.g., carbon steel, stainless steel, brass, etc. (ii) In industry one finds, more often than not, an environment which is a mixture of many bacteria not a monoculture. So, our laboratory experiments should be designed so as to carry experiments in multiculture media in such a way that they can be analyzed correctly and hence be of use in industry. (iii) Work on MIC/MICI performed in laboratories may be mostly done by microbiologist whose biofilm work is on glass/polymer substrate which help in analyzing the details of biofilm and related metabolic reactions through microscopic and other studies. To utilize this information for industry, one needs knowledge of metals and how do they interact with microorganisms in industrial media to get affected through microbial corrosion, etc., which in turn governs the useful life of industrial machinery. For these details, one needs help of a corrosion engineers. So for a work on MIC/MICI to be of use for industrial application so that engineering materials can be used, in real-life situation, as material of construction of industrial machinery where bacteria are to be encountered, it is essential that microbiologists should work in close association with corrosion engineers so as to help each other by making use of each others' knowledge in their respective field.

References

1. Videla HA (2002) Prevention and control of biocorrosion. Int Biodet Biodegrad 49:259–270
2. Videla HA, Herrera LK (2005) Microbiologically influenced corrosion: looking to the future. Int Microbiol 8:169–180
3. Al-Darbi MM, Muntasser ZM, Tango M, Islam MR (2002) Control of microbial corrosion using coatigns and natural additives. Energy Sour 24:1009–1018
4. Zhai X, Ren Y, Wang N, Guan F, Agievich, Duan J, Hou B (2019) Microbial corrosion resistance and antibacterial property of electrodepodited Zn-Ni-Chitosan coatings, Molecules 24(10):1974
5. Guo J, Shaojun Y, Jiang W, Lv L, Liang B, Pehkonen SO (2018) Polymers for combatting Biocorrosion. Front Mater 5:1–15

6. Videla HA, Herrera LK (2009) Understanding Microbial inhibition of corrosion, A comprehensive overview. Int Biodeterior Biodegr 63:896–900
7. Yuan SJ, Pehkonen SO, Ting YP, Neoh KG, Kang ET (2010) Antibacterial Inorganic organic hybrid coatings on stainless steel via consecutive surface initiated atom transfer radical polymerization for biocorrion prevention. Langmuir 26:6728–6736
8. Yang W, Neoh J, Kang KG, Teo ET, Ritsschof SLM (2014) Polymer brush coatings for combating biofouling. Prog Polym Sci 39:1017–1042
9. Yuan S, Tang S, Lv L, Liang B, Choong C, Pehkonen SO (2012) Poly (4-vinylaniline)-polyaniline bilayer-modified stainless steels for the mitigation of biocorrosion by sulfate reducing bacteria in seawater. Ind Eng Chem Res 51:14738–14745
10. Cai W, Wang J, Quan X, Zhao S, Wang Z (2018) Antifouling and anticorrosion properties of one-pot synthesized de-doped bromo-substituted polyaniline and its composite coating. Surf Coat Technol 334:7–18
11. de Carvalho CCCR (2018) Marine biofilms: a successful microbial strategy with economic implications. Front Mar Sci 5:1–11
12. Kip N, van Veen JA (2015) The dual role of microbes in corrosion. Mini Rev ISME J 9:542–551
13. Jayaraman A, Cheng ET, Earthman JC, Wiid TK (1997) Axenic aerobic biofilms inhibit corrosion of SAE 1018 steel through oxygen depletion. Appl Microbiol Biotechnol 48:11–17
14. Jayaraman A, Earthman JC, Wood TK (1997) Corrosion inhibition by aerobic biofilms on SAE 1018 steel 47:62–68
15. Jayaraman A, Örnek D, Duaret DA, Lee CC, Mansfeld FB, Wood TK (1999) Axenic aerobic biofilms inhibit corrosion of copper and aluminum. Appl Microbiol Biotechnol 52:787–790
16. Iverson WP (1987) Microbial corrosion of metals. Adv Appl Microbiol 32:1–36
17. Jayaraman A, Hallock PJ, Carson RM, Lee CC, Mansfeld FB, Wood TK (1999) Inhibiting sulfate reducing bacteria in biofilms on steel with antimicrobial peptides generated in-situ. Appl MIcrobiol Biotech 52:267–275
18. Grooters M, Harneit K, Wöllbrink M, Sand W, Stadler R, Fürbeth W (2007) Novel steel corrosion protection by microbial extracellular polymeric substances (EPS)—biofilm-induced corrosion inhibiton. Adv Mater Res 20–21:375–378
19. Stadler R, Fuerbeth W, Harneit K, Grooters Woellbrink M, Sand W (2008) First evaluation of applicability of microbial extracellular polymeric substances for corrosion protection of metal substrate. Electrochimica Acta 54:91–99
20. Stadler R, Fuerbeth W, Grooters M, Jnaosch C, Kuklinski A, Sand W (2010) Studies on the application of microbially produced polymeric substances as protecting layers against microbially induced corrosion if iron and steel. NACE international corrosion conference, 2010, Texas
21. Zuo R, Örneck D, Wood TK (2005) Aluminum- and mild steel—binding peptides from phage display. Appl Microbiol Biotechnol 68:505–509
22. Zuo R (2007) Biofilms: strategies for metal corrosion inhibition employing microrgansims. Appl Microbiol Biotechnol 76:1245–1253

Printed in the United States
By Bookmasters